OUTBREAK CULTURE

OUTBREAK CULTURE

The Ebola Crisis and the Next Epidemic

PARDIS SABETI *and*
LARA SALAHI

Harvard University Press

Cambridge, Massachusetts | London, England | 2018

Printed in the United States of America

First printing

Library of Congress Cataloging-in-Publication Data

Names: Sabeti, Pardis, 1975– author. | Salahi, Lara, 1986– author.

Title: Outbreak culture : the Ebola crisis and the next epidemic / Pardis Sabeti and Lara Salahi.

Description: Cambridge, Massachusetts : Harvard University Press, 2018. | Includes bibliographical references and index.

Identifiers: LCCN 2018014987 | ISBN 9780674976115 (alk. paper)

Subjects: LCSH: Ebola virus disease—Africa, West—History. | Ebola virus disease—Africa, West—Epidemiology. | Epidemics. | Public health—Africa, West—History.

Classification: LCC RC140.5 .S23 2018 | DDC 616.9/1800966—dc23

LC record available at https://lccn.loc.gov/2018014987

THIS BOOK IS DEDICATED TO

Our families, for their support

Our colleagues, for their collaboration

All who are working to improve the culture of global emergency response

Dr. Sheik Humarr Khan, for all that he gave and all that we learned

CONTENTS

ACRONYMS AND ABBREVIATIONS

CDC	Centers for Disease Control and Prevention
DRC	Democratic Republic of the Congo
ETU	Ebola treatment unit
MERS	Middle East respiratory syndrome
MSF	Médicins sans Frontières (Doctors without Borders)
NGO	Nongovernmental organization
PPE	Personal protective equipment
SARS	Severe acute respiratory syndrome
UN	United Nations
USAID	United States Agency for International Development
USDA	United States Department of Agriculture
WHO	World Health Organization

PREFACE

This book grew out of conversations between Pardis Sabeti and Lara Salahi as a result of our work together on Ebola Deeply, a website that provided measured and constantly updated news coverage during the Ebola epidemic that broke out in West Africa in 2014. Members of the Sabeti Lab at Harvard University, along with many of their colleagues, were involved in the Ebola outbreak from the very beginning. They were direct witnesses to the divisiveness that prevailed among individuals responsible for outbreak response during a time when collaboration was essential. The discord took many forms: individual interests taking precedence over the collective good, miscommunication, time wasted on bureaucratic issues, toxic interactions among people working together under stressful circumstances.

Outbreak culture is the term we use for the collective mindset that can emerge in the beginning stages of a disease outbreak among those who are attempting to respond to it. It forms in the crucible of a chaotic, unpredictable, and potentially fatal environment, in which initial reactions may be based on fear, on the instinct to protect oneself or other people or institutions, or on the desire for exploitation. Outbreak culture dictates how we respond to a pathogen; it often inhibits efficient action and, in some cases, can even perpetuate the epidemic. The pathogen no longer becomes the focus of the outbreak.

All of the disorder during the Ebola outbreak culminated in the painful event that became a metaphor for outbreak culture: the death of Dr. Sheik Humarr Khan, the head of Kenema Government Hospital in Sierra Leone. Khan was a renowned physician with unparalleled expertise in tropical diseases, and Kenema Hospital was one of only two centers in West Africa before the outbreak designed to detect and treat viral hemorrhagic fevers such as Ebola. Sabeti had been working with Khan in 2014 as he and the hospital found themselves at the epicenter of the outbreak in Sierra Leone. Khan should have been able to focus on building Kenema's capacity to handle the surge of Ebola patients. Instead, he faced pressure by individuals who appeared to be seeking to undermine his efforts. Khan was already treating over 80 Ebola patients in a center designed to handle only a dozen, and his calls for help were not heeded. What is devastating is not only that he himself ultimately died from Ebola, but that he died feeling a lack of support. In his final months, Khan experienced firsthand the destructive culture of outbreak response.

We are not the first, nor will we be the last, to tell Khan's story. He has been the subject of a number of profiles during his life and after his death. In addition to consulting those publications, we also relied on firsthand accounts of those who knew and worked with him. We also contacted several of his family members, including some of his brothers and sisters, and his father, Ibrahim Seray Khan, who provided permission to share their stories. This is the first book that has received his father's blessing. Short of writing Khan's biography—which is not our intention—we aim to provide a comprehensive story of his life and work in the context of the Ebola outbreak.

As is the case with any narrative, this book is subject to a number of biases. First and foremost, Pardis Sabeti, as one of the individuals who responded to the Ebola outbreak, is part of the outbreak culture we describe. There is no way to fully disentangle the viewpoint she may have in describing the events that played out. Sabeti directly witnessed how the dysfunction of outbreak culture affected Khan in the final days of his life. We have attempted to place Khan's story in a broader perspective through the use of a survey of over 200 individuals who could offer different versions and opinions of the events. Extensive reporting by Lara Salahi, including numerous interviews, assisted in creating a balanced, yet honest, narrative.

Together, we seek in this book to inform the public about aspects of the environment during disease outbreaks that have little to do with the pathogen itself. Even though our conclusions are broadly applicable, at the same time, our focus is on Ebola. We set out to understand how the people directly engaged in the response to the 2014–2016 Ebola outbreak regarded their involvement. Did other clinical responders, data collectors, and academic researchers experience the same challenges as Sabeti and her team? If so, how did those challenges compare with what they experienced during previous outbreaks? Is it possible to define a culture that has emerged from the global community's response to large infectious disease outbreaks? To date, there are few large-scale published studies on the views of clinical responders, researchers, and health agency personnel concerning the personal and professional challenges they faced during the Ebola outbreak. That is what we wished to explore. Beyond understanding how people felt during the crisis, we wanted to solicit suggestions for change. What ideas did those involved in the outbreak have about how to change attitudes and behavior so that the response to future outbreaks will be more effective, efficient, and transparent?

To better understand the culture of outbreak response and what can be done to improve it, we located as many people as we could who took part in the response to the Ebola outbreak. The more people we communicated with, the more we discovered similarities in negative experiences, challenges, and frustrations, often described as being common to all outbreaks. Some people confirmed the existence of outbreak culture but refused to discuss it on the record, or at all. Some cautioned that placing too much emphasis on the politics of an outbreak would discount the danger of whatever pathogen is at the heart of an epidemic. But many were willing to share their experiences with us directly, in interviews. Others participated via the survey we conducted.

To find individuals who responded to the 2014–2016 Ebola outbreak, we combed through online periodicals, news articles, websites, and published scientific research. We also looked for speakers' names on conference lists and viewed videos of and attended public appearances. We gathered contact information for a thousand people in any part of the world who were involved in one or more of the following areas related to Ebola outbreak response: academic and industry research, diagnostics, drug development, laboratory research, clinical care, clinical research, data collection and analysis, interagency communication, health communication research, health education and public health advocacy, and infrastructure and capacity building. The individuals identified came from Sierra Leone, Liberia, Guinea, Senegal, Nigeria, Mali, Italy, Spain, the United Kingdom, and the United States. We sent an online survey to these individuals, and, in order to broaden the reach of our study, asked them to forward the survey along to anyone they knew who worked in one of the areas we had identified. We received completed surveys from 220 respondents.

The survey included both closed- and open-response questions that inquired about the individual's experience with response efforts during the outbreak, and, when relevant, how their experience compared with previous response experiences. The survey asked them to indicate which of several challenges they faced and invited them to elaborate. Some of the questions were stratified by periods of time or stages of the outbreak, with the stages defined as beginning (March to October 2014), middle (November 2014 to August 2015), and late (September 2015 to the present). This allowed us to identify any trends in perceptions, attitudes, and behaviors over the course of the outbreak. Additional questions focused on solutions to improve future outbreak response. One week before sending the survey, we sent an email outlining the contents of the survey and our goals for conducting the research. Participants were also assured of confidentiality: they would remain unidentifiable, and they would not be linked to their responses. To avoid any opportunity for bias, data analyses were conducted by multiple individuals who were not part of the Sabeti Lab and were not involved in outbreak response.

We received responses from a wide variety of people, though a majority fell into a few broad categories. Over half worked in Sierra Leone in some form of capacity building (e.g., providing supplies, constructing facilities, or creating protocols), were on the ground during the height of the epidemic (spring 2014 to spring 2015), and were employed by international agencies.

A detailed look at the makeup of the respondents shows that they were involved in many different aspects of outbreak response. Respondents defined their roles as the following (more than one role could be indicated):

Infrastructure and capacity builder: 44 percent
Public health advocate: 41 percent
Health educator: 40 percent
Data collector: 34 percent
Clinical responder: 17 percent
Researcher: 14 percent (academia); 2 percent (industry)
Other: 42 percent (e.g., humanitarian worker, epidemiologist, NGO program coordinator)

Sixty-nine percent of respondents worked in Sierra Leone, 35 percent in Liberia, and 22 percent in Guinea. (Other countries in which less than 10 percent of respondents worked were Democratic Republic of the Congo, Ghana, Italy, Mali, Nigeria, Senegal, Spain, and the United States.) Over 50 percent of these workers spent at least 75 percent of their time at sites with active Ebola infections.

Survey respondents reported working for a variety of employers: international agencies (42%), nonprofit organizations (36%), government agencies (21%), academic institutions (12%), and for-profit companies (2%). Most workers tracked the emergence, progression, and transmission of the disease on their own (81%) or from colleagues rather than through their employer.

A vast majority (95%) of respondents reported experiencing challenges while responding to the outbreak. These challenges fell mostly into seven major areas: (1) communication and coordination problems with individuals, organizations, and nations; (2) implications of the delayed response; (3) purposeful lack of transparency and information sharing; (4) resistance by local people; (5) data hoarding; (6) exploitation of power indirectly related to the outbreak (e.g., theft, bribery, corruption, sexual assault); (7) restrictions on research.

These results have some limitations owing to small sample size and selection bias. Our survey of 220 people includes only a fraction of the thousands who responded to the Ebola outbreak and may not be representative of all responders. A majority of the individuals we solicited for the survey did not live in any of the three countries in West Africa where most of the Ebola cases occurred. It is important to remember that an entire community of local health workers also responded to the outbreak. Their experiences are essential to improving outbreak culture. However, an important aspect of the survey is that it was anonymous, so participants were able to respond freely. Open-response questions, which allowed respondents to express their experiences, concerns, and suggestions, were answered by 132 people. The findings provide a glimpse into some shared experiences and challenges of outbreak response.

We supplemented the survey data with in-person, telephone, or email interviews with more than 100 individuals who had firsthand knowledge of at least one part of the international Ebola outbreak response. Some had also been involved in previous large-scale outbreak response efforts. The interviewees included clinical responders, data collectors, scientists, and anthropologists who were on the ground at various stages of the epidemic; foreign and local government officials who were among the decision makers during the response; and local clinicians, Ebola survivors, and victims' family members. Where interviewees recounted conversations, we tried to verify them through additional witnesses. While it is nearly impossible to verify conversations with individuals who are now deceased, direct quotes were used to preserve scenarios recalled by reliable sources. We did not alter any names, but in some cases, we have provided only first names or descriptions of their role during the outbreak. The names of interviewees and dates of interviews are given in the notes.

The emotions that we describe people as feeling were determined through conversations, and in some cases, verified by email and written memoirs. In the case of deceased individuals, we confirmed their emotions with people who had direct contact with them at the time. The overall sentiments as reflected in this book may not represent the emotions of everyone who was involved in the Ebola outbreak response, but we have aimed to accurately reproduce the general atmosphere during a trying time.

Like the survey, these interviews present a somewhat biased sample, made up of the people we contacted and who agreed to be interviewed. Not all interviews are included in this book. We focus on those that told a compelling narrative that we believe represents the greater story. With the exception of survey responses, we did not rely on anonymous sources. If information was given in confidence or on background, we worked to find sources that would confirm it on the record.

While we have included descriptions and narratives of events in Liberia and Guinea, our emphasis is on Sierra Leone, and in particular the crucial events that occurred in Kenema. There were more than twice as many laboratory-confirmed cases of Ebola in Sierra Leone as in Guinea or Liberia. Kenema Government Hospital was initially the only established medical center among the three countries that was able to handle Ebola cases and where active research on viral hemorrhagic fevers was taking place, placing it in a unique position during the outbreak. At the heart of our story is Sheik Humarr Khan. He is officially recognized by Sierra Leone as a national hero and was revered around the world for his expertise on viral hemorrhagic fevers. His legacy remains a part of national infectious disease surveillance and detection programs led by West Africans, which was his hope for the future of outbreak response. Our hope is that readers will connect with Dr. Khan, not just by learning the harrowing details of how he died, but by seeing how he lived for the community he served.

The global community responds to outbreaks with an all-or-nothing approach. Dangerous pathogens are usually ignored until they reach epidemic or pandemic scale. This must change. So, too, must the culture that is formed during an outbreak response. It is not our intention to indulge in that culture by placing blame on or undermining the efforts of any person or organization. Enough of that was done during the outbreak itself (as you will read) and during the months and years following it. However, we must understand the contours of this culture if we are to change the way we approach the next epidemic. We owe it to future responders and to the memory of Dr. Khan.

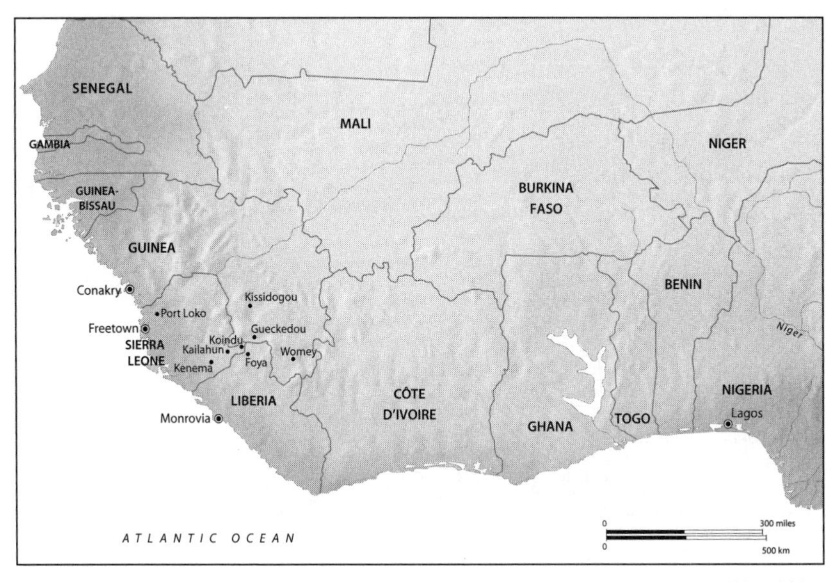
SENEGAL
GAMBIA
GUINEA-BISSAU
GUINEA
MALI
NIGER
BURKINA FASO
BENIN
Niger
NIGERIA
Lagos
TOGO
GHANA
CÔTE D'IVOIRE
LIBERIA
Monrovia
SIERRA LEONE
Conakry
Port Loko
Freetown
Kissidogou
Gueckedou
Koindu
Kailahun
Kenema
Foya
Womey
ATLANTIC OCEAN
0
300 miles
0
500 km

West Africa

PROLOGUE

THE PEOPLE'S FIGHTER

Eventually all shall be well with my people.

—Dr. Sheik Humarr Khan

DR. SHEIK HUMARR KHAN, the head of Sierra Leone's Kenema Government Hospital, sat at his desk and began frantically drafting an email. He was overcome with distress. It was a dry, cloudless day in July 2014, and the concrete buildings with tin roofs that made up the hospital shaded patients and workers from the sun's rays. The near 90-degree temperature kept a shine on his bald head. Beads of sweat collected on his forehead until the drops skidded one by one down each side of his face.

Khan glanced out the corner window to his right, toward the building known as the Lassa ward. His office was located in the administrative building, where the hospital staff regularly gathered, and the ward was its own separate building. The ward had once housed anyone who showed up with signs of hemorrhagic fever—any of several deadly viral diseases that cause fever and

Dr. Sheik Humarr Khan outside Redeemer's University, during the launch of the African Center of Excellence for Genomics of Infectious Disease (ACEGID). *Photo by Pardis Sabeti/© Sabeti Lab.*

bleeding. Kenema was one of only two hospitals in West Africa with the capacity to treat hemorrhagic fevers, and the ward had originally housed patients suffering primarily from one of these diseases, Lassa fever. But now it found itself at full capacity with patients who had a different form of hemorrhagic fever: Ebola.

Kenema, situated in the Eastern Province, is the third largest town in Sierra Leone and the closest to neighboring Liberia and Guinea. Kenema Government Hospital had a 25-bed Lassa ward, with two to four beds in each 11-by-11-foot room. A nearby building contained a specialized diagnostic laboratory where research on viral hemorrhagic fevers was carried out. Fewer than 20 people staffed the ward and lab, including Khan and a group of nurses, laboratory technicians, and surveillance officers.

The previous December, nearly 1,300 miles away in Kailahun, an eastern district of Sierra Leone near the border of Guinea, a traditional village healer claimed that she had the power to cure a mysterious illness that had been striking a growing number of people. Dozens of Guineans infected with Ebola crossed the border to seek help from the healer. She became infected herself and died in May. Many people came to her funeral, where her body was washed in a traditional ritual.

On May 24, a pregnant woman who had attended the healer's funeral arrived at Kenema's maternity ward after suffering a bloody miscarriage. The next day, Augustine Goba, director of the hospital's diagnostic laboratory, ran a series of sensitive molecular tests on a blood sample from her and on a sample from another woman, which had been sent from Kailahun. On May 25, he detected the first known case of Ebola in the country (in the sample from Kailahun), as well as the first case in the hospital (in the sample from the pregnant woman). The results were confirmed by the Sabeti Lab and Nadia Wauquier of the epidemic tracking company Metabiota. On the same day, the WHO was notified that Ebola had spread to Sierra Leone.

Two other women who had attended the healer's funeral arrived at the hospital with fever and chills. Khan was sure they had Ebola, and a diagnostic test proved it. He immediately let his American colleagues know that his team members were bracing themselves for more cases. Robert Garry, a professor of microbiology and immunology at Tulane University, who had been collaborating with Khan, stuffed nine trunks full of supplies, including gloves, gowns, and any additional personal protective equipment (PPE) he could find, and two days later boarded a flight to Sierra Leone.

Khan had isolated the first woman in the Lassa-turned-Ebola ward, where he and the nurses worked around the clock to treat her. She survived, and no one else was infected. Had this woman been one of the first cases of Ebola in West Africa, the hospital's rapid diagnosis and expert handling might have controlled the outbreak. But by May, the epidemic had already been raging for six months, with hundreds of cases in neighboring Liberia and Guinea. When the hospital's outreach team traveled to the patient's village, they found that 14 people had already been infected with two distinct versions of the virus. The number of cases in Sierra Leone kept growing, and the virus would eventually spread to nearly every district of the country.

Beginning in June 2014, dozens of patients started showing up at the hospital with Ebola-like symptoms. Those who had traveled the tree-lined, red dirt road through the gates of the hospital in the past had usually done so because it was well known for its ability to treat hemorrhagic fevers. They would spend a few days in the hospital and then return home. But this time was different. In Sierra Leone, the number of people infected with Ebola virus was doubling each week, and so was the number of deaths. People who expected to stay for days remained for weeks.[1]

Diagnostic labs at Kenema Government Hospital. The hospital contains a biosafety level 3 lab, one of the only labs in West Africa with state-of-the-art diagnostic and research capabilities. These labs were used to test patient samples and confirm Ebola cases during the outbreak. *Photo by Anna Lachenauer/© Sabeti Lab.*

The crowd of Ebola patients entering Kenema was unmanageable. International agencies tasked with responding to the outbreak had already exhausted their limited resources during the previous months, when the epidemic was raging in neighboring Guinea and Liberia. But just as the disease had started to spread into Sierra Leone, cases had been declining in Guinea; the CDC and other agencies believed the outbreak was waning and sent staff home. Kenema Government Hospital, the only place in Sierra Leone prepared to handle cases of hemorrhagic fever, took the majority of patients. While it was well positioned to manage a few cases, or even 20, it was not prepared to take on upwards of a hundred Ebola patients at one time. An addition—with wooden walls and a corrugated tin roof—was built to fit more beds, but the improvements weren't enough. Even the best institutions have a threshold, and that threshold was far exceeded.

The hospital's staff created makeshift areas to separate confirmed cases from suspected cases, as well as from family members and medical staff, who donned and doffed PPE to prevent contamination. The basic design and flow of the ward, however, was a deadly mess. Many who entered the gates never left.

Even before Khan became head of the hospital, the 39-year-old physician was a globally renowned expert in tropical diseases. He had been instrumental in building a thriving center for clinical care and research on Lassa fever at Kenema Government Hospital after the decade-long civil war in Sierra Leone. Before the Ebola outbreak, Khan had treated more cases of hemorrhagic fever than perhaps anyone else in the world.[2]

A medical officer from the World Health Organization (WHO) arrived in Kenema in June 2014 to report on the hospital's operations and to offer help. The hospital had until then had no support from international organizations. Health workers in Kenema had gone on strike over delayed hazard pay, further exacerbating the dire situation. Khan was initially thrilled that calls for more supplies and medical staff were finally being answered. He had spent weeks raising the alarm and asking for reinforcements and support. He was relieved to see clinicians entering the ward and delivering hands-on support. As the visit went on, however, Khan suspected that WHO officials were trying to push him out of his position.

Soon, his fears were confirmed. Immediately after the visit, the officer contacted the Ministry of Health and Sanitation and recommended that the hospital should either be shut down or that control should be transferred to a foreign medical team better equipped to handle emergency cases. The officer believed there were insufficient safety protocols in place to protect the staff from contracting

Ebola. He also recommended that the laboratories, where researchers from Tulane University and the private company Metabiota were carrying out molecular testing of blood samples, should be taken over.[3]

The medical officer recommended stopping what he regarded as on-site Ebola research, on the grounds that the hospital had not received necessary permissions from the country's health ministry. (The hospital and its academic partners had in fact ceased research on site and gained permission to send discarded clinical samples abroad for analysis.) The WHO planned to house its mission at the hospital and provide its own leadership, staff, and resources.[4]

Khan believed the visit was an attempt to undermine him and the entire hospital. Before the outbreak, Khan's position was highly coveted by the West African medical community. Now international responders wanted him replaced. Having already endured many weeks of increased division and distraction by those vying to compromise his position, he was exhausted. By the time he called Dr. Pardis Sabeti of Harvard University for guidance, he was in complete despair. "I don't know why they're doing this," Khan told Sabeti. "We are alone here, and we need help."[5]

Khan decided to write an email to the medical officer, asserting his authority and presenting his point of view. "I believe there is no other person better locally based in this Program than me," he wrote.[6] Khan did recognize that the already dire situation was growing worse. The limited resources were inadequate to manage patients dying from Ebola and to protect his staff. But the officer did not discuss ways to collaboratively improve conditions, something Khan desperately wanted. Nor did the officer address his concerns, some of which Khan felt were unfounded, in person. "I may have then been able to show you that we have already engaged and received ethical approval from the MOHS [Ministry of Health and Sanitation] and from our US partners, and given you more

A group of nurses at Kenema Government Hospital, months before the Ebola outbreak hit. *Photo by Pardis Sabeti/© Sabeti Lab.*

background on our work, before you reached out to others creating unnecessary confusion and concern," Khan wrote.

The hospital staff in Kenema were well positioned to detect and treat Ebola because of their experience with other tropical diseases, such as malaria, yellow fever, tuberculosis, and Lassa fever. Since the 1990s, Kenema had maintained Sierra Leone's only treatment center for viral hemorrhagic fevers. Dr. Aniru Conteh, the world's leading Lassa fever specialist, headed Kenema's operation after it moved from Nixon Memorial Hospital in Segbwema, about 60 miles west. Conteh treated thousands of patients in Kenema's Lassa ward during Sierra Leone's civil war. In 2004, Conteh was pricked by a needle while treating a patient and became infected with the Lassa virus. He died less than three weeks later from the disease he had dedicated his life trying to eradicate.

Khan, who had completed his internal medicine residency in Ghana less than a year before Conteh's death, applied for the vacancy and was hired immediately. Overseeing the hospital and its unique hemorrhagic fever program was a highly sought-after position. Khan was thrilled about the appointment, but one of his brothers, who lived in the United States, did not feel the same way. "We begged him to come to the United States and practice here," Khan's elder brother Sahid said. "He would tell us, 'There are too many doctors there. There are not enough in Sierra Leone.'"[7]

After finishing his residency, Khan, who had grown up in Sierra Leone, traveled widely to learn everything he could about tropical diseases, with the goal of bringing the knowledge home. He attended medical meetings all over the world and collaborated with other researchers on publications. Before the Ebola outbreak emerged, he had planned to conduct research on tropical diseases with Sabeti at Harvard University.

Khan shared with his colleagues an interest in eradicating infectious diseases and training the next generation of doctors and nurses. In May 2014, he met with dozens of public health experts from across West Africa and the United States at Redeemer's University in Lagos, Nigeria. The group, known as the Viral Hemorrhagic Fever Consortium, included Robert Garry of Tulane University, Pardis Sabeti, and Nathan Yozwiak, then a project manager in the Sabeti Lab. Using grant money from the World Bank, the consortium launched a program at Redeemer's University called the African Center of Excellence for Genomics of Infectious Diseases. The venture would provide West Africans with education, training, and resources to enable them to monitor, study, and treat dangerous pathogens in their own region.

Part of the initiative involved establishing a state-of-the-art Lassa ward in Kenema, where work on the Lassa virus was already underway. Lassa fever, endemic to West Africa, is one of the most common hemorrhagic fevers in the region. In some areas of Sierra Leone and Liberia, 10–16 percent of people admitted to hospitals every year are diagnosed with the disease.[8] It was often the first diagnosis that Khan considered when a patient came into an exam room in Kenema complaining of pain, fever, and nausea.

Members of the consortium strategized about how to create a program that would support and educate West African scientists and clinical workers so that when the next outbreak occurred, they would be better able to handle it than any outside agency. Long days spent in a conference room were often followed by long nights decompressing in Yozwiak's hotel room. Redeemer's University, a Christian institution, was a dry campus, but most nights, the newly formed team would pay a driver to buy them enough spirits to last the evening. They gathered in Yozwiak's room—which they called "Nathan's Speakeasy"—for beers and conversation. Happy hour lasted well into the night. "We all felt this was the beginning of a great and long journey together," Yozwiak said, remarking on the camaraderie in the room.[9]

During one evening discussion, Khan talked about Sierra Leone's civil war. He discussed his dream of making highly contagious and untreatable viral diseases survivable. Lassa fever had been his life mission, but Ebola now preoccupied him. Khan revealed that there was nothing he feared more than treating Ebola-infected patients. "I'd rather get Lassa than Ebola," he told Yozwiak. Lassa is no less horrific than Ebola. The two diseases have similar symptoms, beginning with fever, muscle aches, weakness, and nausea, and progressing, for Lassa, to tremors, facial swelling, and encephalitis, and for Ebola, to diarrhea, vomiting, and bleeding. Perhaps Khan was thinking of the much higher fatality rate of Ebola, but Yozwiak believed that there might have been something other than a clinical reason behind his fear.

While the team developed plans for the future, unbeknownst to them, Ebola was already beginning its deadly journey toward Kenema.

In May 2014, soon after Khan returned to Kenema from Nigeria, Ebola had already been brewing in Sierra Leone, and the hospital saw its first case. Just 50 yards away from Kenema Hospital's Lassa ward, on hospital grounds, stood the shell of what was to be a much larger and more advanced ward. Construction—which was sponsored in part by the European Union, the WHO, and the US Department of Defense through Navy contractors—would begin, and then stall. As Tulane University virologist Daniel Bausch explained, "Contractors had insufficient capital, budgets were frozen, funds were lost, and key personnel changed."[10] A contractor once promised the Kenema staff that a railing would be built at the entrance of the building, but the Navy didn't see the railing in its construction plans. This discrepancy stopped construction for months. "If the ward had been built on schedule and as planned, they would've had this big hospital that could have handled all of [the Ebola cases]," Yozwiak said.

International agencies, including the WHO and Médicins sans Frontières (MSF), had launched a response to the outbreaks in Liberia and Guinea, but had few resources left to spare for Sierra Leone. The US Centers for Disease Control and Prevention (CDC) had yet to initiate an emergency response. Garry and Sabeti's research group reached out to numerous agencies, pleading for resources and support to be directed to Kenema. "We are working to advocate for you all, but we are not sure we have a say ourselves," Sabeti wrote in an email to Khan, recognizing that they had little sway over the larger agencies.[11]

The WHO—which had been responding to calls for assistance in other parts of Sierra Leone—sent an official to Kenema Hospital to assess its capacity to handle cases, and to see where the agency could best establish its operations. It was then, in June 2014, that Khan first met with the medical officer who questioned the hospital's ability to respond to the epidemic. "If global responders react to an outbreak and it's in a vacuum, then they can set up whatever operation they want; but it's never in a vacuum," Yozwiak said. "They're entering certain scenarios in which there are preexisting expertise and authority and capabilities. They come in and often try to collaborate, but they can also be pretty heavy-handed."

In this case, the WHO's plan was to supervise the clinical care of Ebola patients. A team under the agency would be involved in every aspect of care, from the minute someone running a fever entered the hospital. The agency had its own protocol that the local nurses and staff were required to adopt. The agency would also approve any testing and research that was taking place on site. Local physicians and healthcare workers regarded the agency's plan less as forming a partnership than as taking full control of hospital operations and personnel. Since Kenema was a government hospital, the plan had to be signed off on by Sierra Leone's Ministry of Health and Sanitation. That had already been done.

As Ebola cases multiplied in Kenema, Khan and the nurses spent more hours in what became the Ebola treatment unit (ETU) than their dying patients did. The smell of vomit, blood, and death pervaded the narrow stifling hallway and the rooms devoted to Ebola patients. Using a tarp and stakes, the hospital added patient holding centers to triage those requiring immediate attention. The hospital was composed of mostly unlabeled ranch buildings, connected by outdoor concrete paths that overlooked the rolling hills of the countryside. The ETU was an unassuming building among them. By June 2014, people knew exactly which building to enter.

Khan relied on his nurses to maintain organization in the chaotic ward. He and his staff were treating up to 80 patients at a time, working 16-hour days. A rigorous decontamination procedure had been established. Staffers wore PPE consisting of heavy full-body suits, face shields, boots, and outer gloves. After treating patients, they were sprayed with a chlorine solution before undressing according to a carefully prescribed procedure.

Still, that wasn't enough. Nurses were being infected at an alarming rate. Khan was in desperate need of additional protective gear and medications. His pleas increased when his staff began dying from the very virus they were treating. He reached out to anyone he thought would listen. In July, he sent to Sabeti a draft of the letter he was planning to send to the medical officer who had taken charge of the unit: "Please may I remind you that your objective here is to help me manage our patients, and on that note please allow me to perform my administrative role. I actually requested for a clinician to come and help me manage our cases because I am alone, and if this is the case I expected whosoever comes should [seek me out with] any concerns that may arise in my unit first before any other body because that person might not . . . know the true picture . . . here, and if this proves bigger than me then I know the appropriate [administrative] steps to take."[12]

The outbreak, indeed, was bigger than him.

In recent days, Khan had felt like the sole combatant in an all-out war. He fought to protect his staff and all of Sierra Leone against Ebola's presence in the region. He fought to keep Sierra Leoneans and those who came to the hospital alive. As the epidemic expanded, large relief agencies sent representatives to assess its scope. Groups of scientists entered the region to gather virus samples and surveillance data. Even before the outbreak, members of Sierra Leone's health ministry had been eyeing Khan's influential position as head physician in a government hospital that could handle viral hemorrhagic fever cases. Khan felt he was spending more time fighting outside agencies than fighting the virus.

Khan's email to the officer concluded, "In any case in the future please kindly work with me . . . so that eventually all shall be well with my people."

He sent the email to Sabeti to review after discussing his many concerns with her by phone. Within hours, Sabeti sent a response. "You are the leader here," she wrote to Khan. "And I am hopeful he will understand that he is here to support you."

By July 2014, the virus had fully occupied Kenema District. One of every two people infected with Ebola would die. Khan watched nearly his entire nursing staff succumb to the disease. One nurse, Alex Moigboi, began feeling ill after a long day of caring for Ebola patients. Khan examined Moigboi without wearing PPE. He checked Moigboi's eyes and touched his skin. Khan thought it was malaria. He ordered Moigboi to take a blood test and go home to rest. Moigboi's blood test came back positive for Ebola.

It is likely that Khan's examination of Moigboi led to his own infection, though it is impossible to know for sure. The morning after Moigboi's death in Kenema Hospital, Khan awoke with a fever. His assistant, Peter Kaima, thought it was a combination of stress from hours of clinical care and dwindling staff. Khan knew it was more than stress. The first blood test came back negative for Ebola. Another test, taken 24 hours later, on July 21, confirmed his fear: he had Ebola.

Simbirie Jalloh, program coordinator for Kenema's Lassa Fever Research Program, called Robert Garry to deliver the news. "Things were spiraling out of control at that time," Garry recalled.[13]

As Khan battled for his life, government medical officials feared that news of his illness would cause hysteria among townspeople and patients in the ward. Though the virus permeated every inch of the ward and had felled many staff members, Khan had been considered untouchable. Officials initially considered keeping his illness a secret, but it was not possible at that point because the staff had

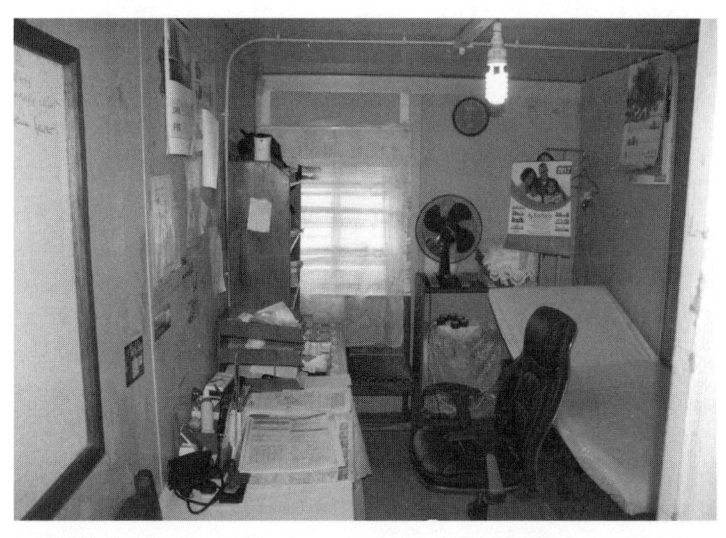

Dr. Sheikh Humarr Khan's office, located near the old Lassa ward, in July 2017. The office is now occupied by Dr. Khan's successor, Dr. Donald Grant. *Photo by Anna Lachenauer/© Sabeti Lab.*

already been grappling with the news. Instead, officials advised that Khan be moved to an MSF-operated center about 75 miles north in Kailahun. Even though his staff begged him to stay in Kenema, Khan agreed to be cared for in Kailahun. "He realized his presence there was demoralizing to the nurses and other staff," Garry said.[14]

On July 23, an ambulance took Khan on the four-hour trip along a muddy dirt road to Kailahun.[15] Health workers anticipating his arrival met him at the center's entrance, but Khan insisted on walking into the isolation ward on his own. He even administered his own intravenous fluids, since MSF's standard protocol for Ebola only called for oral treatments—acetaminophen for pain, antibiotics, and rehydration salts.

Meanwhile, a Sierra Leone government official sent an email to foreign agencies asking about experimental Ebola treatments for Khan. Sabeti and Garry worked to amplify the request. They joined a conference call with a little more than a dozen people, including members of global health programs at the National Institutes of Health, WHO, and CDC, to discuss the staff infections at Kenema and to explore ways to treat Khan, including administering an experimental drug called ZMapp.

ZMapp, a drug containing humanized antibodies from mice that have been exposed to the Ebola virus, had never been used on humans. US and Canadian researchers first tested ZMapp on monkeys infected with Ebola in April 2014, three months before Khan fell ill. Early trials at the Public Health Agency of Canada's research facility in Winnipeg showed that three doses of the serum could knock out the Ebola virus, even when the monkeys seemed close to death. All 18 monkeys that were injected with ZMapp, as many as five days after infection, survived.[16]

In June, Gary Kobinger, one of the Canadian researchers, took three vials of ZMapp from his lab in Winnipeg, traveled to Kailahun, and put them in the freezer there.[17] Kobinger wanted to see how the drug would withstand a tropical climate and poor storage conditions. The vials remained in the freezer, and, at the time of Khan's hospitalization, were almost within his reach.

The decision of whether to give ZMapp to Khan was also discussed among officials in Sierra Leone and within Kailahun's ETU. Senior representatives of governmental and medical agencies in Kenema and Kailahun convened multiple times to discuss the complex issue. No one took the decision lightly. Healthcare workers directly caring for Khan advocated for him to be treated with ZMapp. The WHO clinical leader in Kenema, Dr. Tim O'Dempsey, contacted Kobinger to discuss whether Khan would be a good candidate. Both agreed that it should be tried.[18]

Back in the United States, Sabeti and Garry were feeling hopeful after the conference call. "We were trying to convince them [the agency representatives] that Khan was the best person to give ZMapp," Garry recalled. Others involved in the call described it as "exploratory," with no decisions made on Khan's care. Still, Garry recalled that a WHO lead physician on the call "told us there were doses of ZMapp at his bedside and that he was about to get them or already got them."

Although there were concerns about use of the experimental drug, since no human trials had yet been done, the consensus of participants in the conference call was that the potential benefits outweighed the risks. No one understood those risks better than Khan. Sabeti and Garry argued that Khan himself was in the best position to decide whether to try the experimental drug. Every person on the call believed that all possible attempts should be made to save Khan. After the call, Sabeti drafted a joint statement from the group insisting that experimental therapies be tested if the outbreak persisted.

Khan never received the injection. Unbeknownst to Sabeti and Garry, MSF, the agency in charge of the Kailahun ETU, had made the difficult decision not to give the experimental treatment to him, fearing that if it didn't work, Sierra Leoneans would blame foreign responders for the prominent doctor's death. Another fear was that the public would think foreigners were conducting risky experiments on Africans. And finally, they considered it unethical to single out Khan for exceptional treatment.[19]

Khan was never asked whether he wanted to receive the experimental drug. Many who knew him believe that, if asked, he would have agreed to it without hesitation. "Dr. Khan was educated enough in medicine to give proper informed consent. And I imagine that he, provided the choice, would have volunteered," Yozwiak said. "That's the key thing with experimental treatment, and that, to me, alleviates a lot of the concerns about the stigma of doing research in vulnerable populations. Dr. Khan was not a member of a vulnerable population. He was a distinguished scientist in his community."

Khan lived out the Ebola virus cycle away from his beloved staff at Kenema, though he kept in touch through regular phone calls. During one of these calls, Simbirie Jalloh asked, "Dr. Khan, what is going on? Why can't you come back to Kenema? Here you have people you know very well who can treat you." Khan replied, "Simbirie, this is the conversation I am having now, but they promised me they would take me out of the country for better treatment."[20]

The health ministry was trying to make plans to evacuate Khan by medical flight to a better equipped hospital in Europe or the United States. Air evacuation of an Ebola patient had not been done before. Most countries rejected the ministry's pleas, but officials in Germany finally agreed. By some accounts, the WHO arranged for an air ambulance, operated by International SOS, to evacuate Khan.

There are conflicting reports as to how close Khan came to being evacuated. Some people say that an ambulance sped to Kailahun to transport Khan to the airport, but that International SOS refused to take him because they were not equipped to handle symptomatic Ebola patients—Khan had by that time developed severe diarrhea and was vomiting. (International SOS refused to discuss specific patient evacuations.) Others contend that there were no plans to evacuate Khan. In the initial days after Khan's diagnosis, WHO clinicians working in Kenema asked the agency's field coordinator whether Khan would be evacuated from Sierra Leone. The field coordinator reportedly said that Khan was not eligible for evacuation since he was not a WHO employee.[21] Some believe that this policy influenced the final decision.

Khan himself believed he would be evacuated. He reassured Jalloh that he would be leaving the country for better care. Jalloh sent a suitcase of Khan's belongings, including his passport, with nurse Michael Gbakie, who drove a van from Kenema to Kailahun.

In Kailahun, health workers barred Gbakie from seeing Khan since he was not an authorized medical professional there. For days, Gbakie waited outside the ward while receiving updates from Jalloh on Khan's dire condition. Finally, he barged into a hospital staff meeting and demanded to see Khan.

"Please, this man is my boss," Gbakie said. "He is suffering. Please, allow me to enter the center so I can see him."

"Okay," a nurse told him, pointing to the PPE. "But please don't tell the others."

When Gbakie entered the room, Khan was lying alone on a cot. A nearly empty IV fluid bag hung next to him, with the line still running from his arm. He was soaked in his own waste, crying out for help. Gbakie stormed back outside and ripped off his mask. He was seething.

"Dr. Khan is dying over there," he shouted to healthcare workers. "Please go and pay attention to this guy!"

Gbakie secured his mask and reentered the room. He cleaned Khan and changed his clothes, placing a mat down outside so Khan could lie in the fresh air.

"I'm helpless," Khan told Gbakie. "They are not taking care of me at all." Khan said he was taking his own temperature and changing his own IV. His fever came and went, and so did his ability to stand up and walk. Jalloh sent food from Kenema regularly after she learned he was being given only water and fruit jelly.

"We have to bring him back to Kenema," Gbakie told Jalloh during a phone call one evening. "He is at the point of death."

Toward the end of his life, Khan was aware of the discussions surrounding his care. He was never asked for his input, nor was he involved in the decision making. Some of the staff in Kenema believe that he knew that Kailahun housed the experimental

ZMapp, and as days went by, without being given much more than replacement fluids, he tried his best to reassure his colleagues back home.

"Don't worry, they're going to send me to another country," said Khan one day. "The plane is coming tomorrow."

"No, Dr. Khan, this thing is getting really serious," Jalloh responded.

The next day, Jalloh held a staff meeting to discuss the influx of new cases in their own ward. They also began plotting how to bring Khan home.

It was shortly after 2 P.M. on July 29 when her phone rang. Gbakie was on the other end.

An ambulance drove the four-hour stretch of winding and unpaved road from Kenema to pick up Khan's body and only one other item: his suitcase, which had remained untouched in his room. Nothing else in Kailahun belonged to Khan.

Garry was home in New Orleans when his cell phone rang late in the evening. He heard Jalloh's voice on the other end.

The next morning, Garry emailed Sabeti: "More unthinkable sadness—he is dead."

1

SETTING FOR DISASTER

It was like going into a war zone but you don't know where the enemy is coming from. The enemy was all around you.

—Nahid Bhadelia, National Emerging Infectious Diseases Laboratory, Boston University

THE EBOLA OUTBREAK OF 2014–2016 was unlike any of the 25 previous outbreaks that had occurred in Africa since 1976. The disease spread quickly into urban areas and killed more people than all previous outbreaks combined.[1] It occurred in a region where there is profound distrust of government and where the level of basic healthcare is among the lowest in the world. Poor road conditions, especially in rural areas, where paths are narrow and unpaved, made it difficult to transport aid, equipment, and personnel. Some roads had recently been improved to allow easier access to once remote areas, but this proved to be a double-edged sword, since it also helped to ferry the virus to larger, more populated areas.

A number of other factors also contributed to Ebola's rapid spread and hampered containment during this outbreak, including certain religious and cultural practices and widespread petty corruption.[2] For example, families sometimes bribed burial workers to create certificates saying their loved ones had not died of Ebola so their bodies could be released for a traditional burial.

The region's political history also provided an ideal environment for the outbreak. In 2014, Sierra Leone was still emerging from the ashes of a deadly civil war. More than fifty thousand people had been killed, thousands more were raped or had limbs amputated, and over a million were displaced. The decade-long civil war began in March 1991, when rebels with the Sierra Leonean antigovernment militia the Revolutionary United Front (RUF), assisted by the National Patriotic Front of Liberia, crossed from Liberia into Sierra Leone to attack the town of Koindu. This town would later be one of the key launching points for the Ebola epidemic. The invasion, coupled with the dysfunction of Sierra Leone's army, resulted in the overthrow of the civilian government of Joseph Saidu Momoh. The RUF took control of large areas of eastern and southern Sierra Leone that were rich in diamond mines. Rebels used revenue from the sale of diamonds to finance the insurgency; many Sierra Leoneans believe that control of this precious resource, and not political gain, was at the heart of the revolt.

By the end of 1993, the Sierra Leone Army had made some gains against the RUF. The insurrection appeared to have been contained in 1995, after the government of Sierra Leone hired Executive Outcomes, a South Africa–based private military company. A year later, Sierra Leone's newly elected civilian government brokered a peace deal with the RUF.[3] The government terminated its contract with Executive Outcomes under UN pressure to transition to its own peacekeeping force, but this replacement force proved ineffective. In 1997, before provisions from the peace deal could be implemented, rebel army officers staged a coup and teamed up with the RUF to establish the Armed Forces Revolutionary Council

(AFRC) as the new government of Sierra Leone. The war raged on as attempts were made to restore the civilian government. By that point, international health groups had pulled out of Sierra Leone and neighboring countries.

Nearly a decade before the war began, eight-year-old Sheik Humarr Khan moved from his parents' home in the coastal town of Lungi to Freetown to live with his older brother Sahid.[4] Their father, Ibrahim Seray Khan, had sent Humarr—as he was called—to live with Sahid to take advantage of educational opportunities in the city and learn from his older brother how to be responsible. Ibrahim, a highly respected school headmaster, expected excellence in school from all 10 of his children and urged them to engage in community service.

Sahid was designated to be Humarr's protector until he reached puberty. It was 1983, and Sahid, who had graduated from the University of Sierra Leone with a journalism degree, found his career options were limited to state-run media. He and his friends had pulled together their resources to start one of the first independent newspapers in the country. The paper, *For Di Pipul,* which means "for the masses" in Krio (Sierra Leone's creole language) was a source of pride for the Khan family. It covered everything from local news stories to exposés on human rights abuses in the country. "We were not looking for a revolution," said Sahid. "All we were advocating for was for the government to adhere to the basic human and democratic rights."

As opposition to the dictatorship intensified and the government was losing power, *For Di Pipul* placed Sahid and his two friends in the crosshairs of the military police. Humarr watched as police officers forced themselves into Sahid's apartment with death threats and accusations of treason. He was in awe of his brother's ability to create waves with his publication. "[Humarr] started memorizing the dictionary because he knew the power of words," Sahid said. "He saw they had consequences."

Sahid and his colleagues were detained and released without charge on numerous occasions and resorted to a tactic of publish, distribute, go underground. Humarr, who was 10 at the time, decided then that he, too, would be a journalist. Memorization was easy for Humarr, who had learned the entire Quran through repeated recitation. He could not read or write the Arabic verses, nor could he understand much of what he was saying, but the syllables of each word fell where they should. Humarr did not see the dangers of Sahid's words, but the boys' father did.

Humarr's weekend trips home to Lungi changed his mind about being a journalist. Ibrahim had opened the family home to patients and their families who traveled from across the country to visit nearby Bai Bureh Memorial Hospital. The 50-bed clinic had been founded by a wealthy European woman and her husband, a Sierra Leonean obstetrician-gynecologist who was known in town as "the surgeon." Humarr grew curious about the surgeon and the clinic that drew in so many patients for treatment, including Sierra Leone's president, Siaka Stevens. During his visits home, Humarr often shadowed the surgeon at the hospital. Ibrahim reveled in the idea that one of his children might become a physician.

When it became clear that war was breaking out, Humarr was sent to live with his older sister Isha in Makeni and attend secondary school. By then, Humarr had his sights set on going to medical school. He graduated with near perfect grades, but he faced a serious obstacle to admission.

"At that time, if your family didn't have money, they [the university] would never let you go," Humarr's sister Mariama Lahai recalled. "The first time [he applied] he was not allowed," she said. "My father encouraged him to do something else."[5]

Ibrahim feared his modest salary would keep his son from attending medical school. On a trip to Freetown to buy his wife an airplane ticket, Ibrahim ran into a former student who had gone on to work in a powerful government position. The student recognized Ibrahim and wanted to purchase the ticket for him. Ibrahim refused, asking if the student could help his son get into medical school instead.

It is unclear what, if anything, the student did to tip the scale, but Humarr went on to attend medical school outside Freetown. Meanwhile, Ibrahim had his other children to worry about. Afraid for Sahid's life, Ibrahim urged him to shut down *For Di Pipul* and leave Sierra Leone.

Sahid moved to the United States in 1991 with one of his sisters in time to escape the war. Years later, at the height of the war, a majority of the Khan family, including Humarr, temporarily fled to neighboring Guinea.

A wave of looting, rape, and murder followed the announcement of a new government under the joint forces of the RUF and the AFRC in 1997. The civil war pitted neighbor against neighbor within villages. Unlike the 1994 Rwandan genocide, where the segregating factor was ethnicity (with Hutus targeting Tutsis), in Sierra Leone, differences revolved around political loyalties—those who supported the RUF uprising against those who remained loyal to the government. Political or cultural identity is much harder to distinguish than race or ethnicity, and viewpoints can change. Mistrust and paranoia became just as much a wound of war as the physical scars of brutality.[6]

Internal conflicts also took hold in the neighboring countries of Liberia and Guinea. Civil wars, military coups, and violence destroyed healthcare systems and infrastructure. Many people steered clear of government-run health facilities not only out of mistrust, but because they generally did not have adequate staff or equipment. Alternative treatments, such as self-medicating or using traditional healers, became commonplace.

In 2000, the British government intervened, sending in troops that drove out the RUF and handed control of Freetown back to the elected government. The government established a task force to restore social cohesion, but these peacebuilding efforts did not accomplish much. By 2007, millions of dollars had gone into a special court that prosecuted only nine men for war crimes. The Truth and Reconciliation Committee—a government-mandated task force founded to unearth the details of what took place—never reached beyond the district capitals to ordinary villagers most affected by the war.

The war also resulted in greatly increased rates of mental illness. By war's end, more than 12 percent of people in Sierra Leone were living with mental health disorders, including schizophrenia, depression, anxiety, and substance abuse. This figure is likely a gross underestimate, since stigma kept many families from seeking help. In 2008, six years after the end of the war, an estimated 40 percent of Liberians had symptoms of major depression, and 44 percent appeared to have post-traumatic stress disorder. Despite the overwhelming need, Sierra Leone and Liberia each had only one psychiatrist.[7] In Sierra Leone, the country's sole psychiatrist, Edward Nahim, a Soviet-trained physician in his seventies, worked out of a nineteenth-century psychiatric hospital located on a hill above the capital, Freetown. The hospital, commonly called the "crazy yard," lacked basic sanitation, and patients were sometimes restrained in chains. Stories circulated that even rebel fighters were afraid to enter the wards.[8]

In the years following the civil war, large international organizations poured into Sierra Leone with assurances that they would assist in healing the emotional wounds. But the undertaking, while initially helpful, was temporary. Healing invisible wounds became a tough sell to international donors; the dollars dwindled, and the number of aid workers declined. The burden of mental health care fell on smaller organizations that had limited resources but drew strength from deep connections within the community.

In Sierra Leone, organizations like the US-based Fambul Tok found that the very people and places most ravaged by war can also be the most powerful resources in the painstaking work of rebuilding lives and communities. Yet the people most affected by war are often the least consulted by the international community. Fambul Tok—"Family Talk" in Krio—was a less formal and more community-oriented version of the truth and reconciliation commissions that had been used in other countries after a conflict. The group, founded in 2007, organized truth-telling circles within the villages, where victims bore witness to the atrocities committed against them and perpetrators listened to the victims' accounts and asked for forgiveness. The community then engaged in a cleansing ceremony, and the perpetrator would compensate the victim with a small token.[9]

Fambul Tok's truth and reconciliation forums drew from Sierra Leone's tradition of discussing and resolving issues within the family circle, but expanded it to include an entire community. The process seemed to be working. Nearly a decade after the end of Sierra Leone's civil war, people who participated in the forums began forgiving and trusting one another. New friendships began to develop within communities. By 2013, a wounded society was beginning to heal.[10]

Meanwhile, deep in the Guinean forest, a playful two-year-old boy disturbed a bat-infested tree, unaware that at least one of the bats he came in contact with carried a virus that would propel Sierra Leone and its neighboring countries into the next war. It was December 2013, and the boy, Emile Ouamouno, became ill; he developed a fever, began vomiting, and had black stools. He died four days later in his village, Meliandou, located near the border of Sierra Leone and Liberia. A short time after the toddler's death, his four-year-old sister began experiencing the same symptoms and died, followed by his mother and grandmother. Then, health workers who cared for him and his family began to die, as did others in the village. People in nearby towns began to suffer from the same mystery illness. In March 2014, after three months of investigating, Médicins sans Frontières (MSF) arranged for blood samples to be flown from Guinea to the Pasteur Institute in France. There, researchers discovered that a virus known as Ebola was the culprit. By that time at least 29 people had died and 49 had been infected in three different districts in the country. Although it is impossible to know with complete certainty that Emile was the first victim—or that a bat was the culprit—it is almost certain that the epidemic started with a single infection.[11]

Time, place, people, and the environment all combined to create the perfect stage for Ebola to emerge and spread. Ebola is transmitted through human-to-human transmission, and its spread was facilitated by travel, as well as the lack of clean water, sanitation, and health systems. Ebola spread more quickly in 2014 than in previous epidemics because of the greater interconnectedness of the region, combined with population and infrastructure growth. One of the first larger towns it spread to was on a major trading route between Guinea, Sierra Leone, Côte d'Ivoire, and Liberia that had no health facilities or running water. The population of Africa had grown from 427 million in 1976 to over 1.1 billion by 2014. More roads cut through forests and villages, enabling people to move through towns and cities and providing more hosts for otherwise isolated viruses like Ebola.[12] Deforestation destroyed the natural habitats of animals that harbor rare diseases, driving them into populated communities.

As early as March 2014, responders from the World Health Organization (WHO) known as contact tracers, tasked with identifying and diagnosing people who may have come in contact with an infected person, along with investigators from Guinea's Ministry of Health, had already documented two suspected cases of Ebola in Sierra Leone. The cases involved two women from Kpondu, Sierra Leone, a small village in the Kailahun District, about three miles from the border of Guinea. One of the women, 37-year-old Sia Wanda Koniono, had begun experiencing Ebola-like symptoms after having traveled to Guinea. She returned to Guinea to seek medical treatment, where she died. Later, Koniono's daughter came down with the same symptoms of fever, vomiting, and diarrhea.[13] Within weeks, additional suspected (and later confirmed) cases of Ebola spread to nearby Koindu. The small town where Sierra Leone's civil war had started quickly became one of the main sites of Ebola's spillover into the country. The two suspected Ebola cases, while noted by Guinea's Ministry of Health and the WHO, were not officially reported to officials in Sierra Leone until May. The two months during which the contagion spread without monitoring proved to be a critical time in the outbreak's conversion to an epidemic.

The open border between Sierra Leone and Guinea made it easy for infected people to travel to and from both countries. Many had flocked to Sierra Leone from Guinea after hearing that a traditional healer had special powers to cure the mysterious disease. The healer, an elderly woman named Finda Nyumu, ended up becoming infected and dying, setting off a chain of transmissions that led to an estimated 365 deaths.

On May 25, Kenema Government Hospital, which contained the only laboratory in the country equipped for Ebola testing, confirmed the first recorded Ebola case in Sierra Leone. Two days before, a woman named Victoria Yillah had been admitted to Kenema's maternity ward after suffering a bloody miscarriage, and samples were taken for testing. On the same day, Mamie Lebbie, who lived in a small town in the Kailahun District, became ill. She had probably been infected by her mother-in-law, who had been among the group of people who washed the healer's body at her funeral. A sample of Lebbie's blood was sent to Kenema for testing. Augustine Goba, director of the hospital's Lassa fever laboratory, confirmed that Lebbie was infected with Ebola. She is considered the first laboratory-confirmed case of Ebola in the country, and Yillah is the second. Both women survived.[14]

By sequencing the genomes of the virus from samples taken from Yillah, Lebbie, and others, scientists, including the team headed by Humarr Khan and Pardis Sabeti, found two different versions of the virus circulating among at least a dozen women who had attended the healer's funeral.[15] This meant that not all of the Ebola cases had originated directly from the healer. At least one other person at the funeral must have already been infected and set off a separate chain of transmissions.

The viral species that caused the West African epidemic is *Zaire ebolavirus,* one of five known species within the Ebola genus. *Zaire ebolavirus* has the highest fatality rate and is the most infectious to humans. Never before had this species been transmitted among people for such a sustained period of time.[16]

One of the troubling aspects of *Zaire ebolavirus* is its tendency to sporadically appear and then disappear from human populations. In that sense it is different from other viruses, like measles or chickenpox, which remain among the population at a low frequency after an outbreak. After the first recognized outbreak of Ebola in 1976, near the Ebola River in what was then Zaire and is now the Democratic Republic of the Congo (DRC), the virus vanished for 20 years, only to reemerge in 1995 in the city of Kikwit, DRC. During the 1976 outbreak, epidemiologists worked tirelessly to understand the origin of the virus, to help predict and prevent the next outbreak. They tested hundreds of animals from dozens of species but came up empty-handed. They tried again during the 1995 outbreak, this time testing more than three thousand animals. Gorillas and chimpanzees were dying at high rates from Ebola, but researchers did not believe that the virus had originated in these animals. Researchers also found Ebola antibodies in fruit bats, which many scientists now think are the most likely natural host of the virus. Since contact between bats and humans is rare, so too is an outbreak.

The familiar characteristic of appearance and disappearance also occurred during the 2014–2016 outbreak. A sharp decline in newly reported cases in April and May 2014, especially in Guinea, led responders to believe that the outbreak had ended as quickly as it had begun.[17] In May, the decline in cases in Macenta, Guinea, considered to be the country's epicenter, led to the closure of the MSF treatment facility there. That decline was short-lived.

The virus was still circulating and changing. Each infection gave the virus an opportunity to mutate.[18] While the probability of any particular mutation appearing is exceedingly rare, the Ebola virus had many opportunities to move from human to human, and a mutation eventually did occur that dramatically changed the biology of the virus.[19] This mutation made the virus more effective at entering and infecting the cells of humans and other primates, and less effective at doing so in the cells of bats and other nonprimate mammals. It is likely that this mutation occurred between March and May 2014—during the period of time when cases were identified on the border of Sierra Leone but were not officially reported—just as the trajectory of the outbreak dramatically escalated. Viruses containing this mutation were so successful that over 90 percent of the virus particles analyzed later in the outbreak were found to be descended from this lineage.

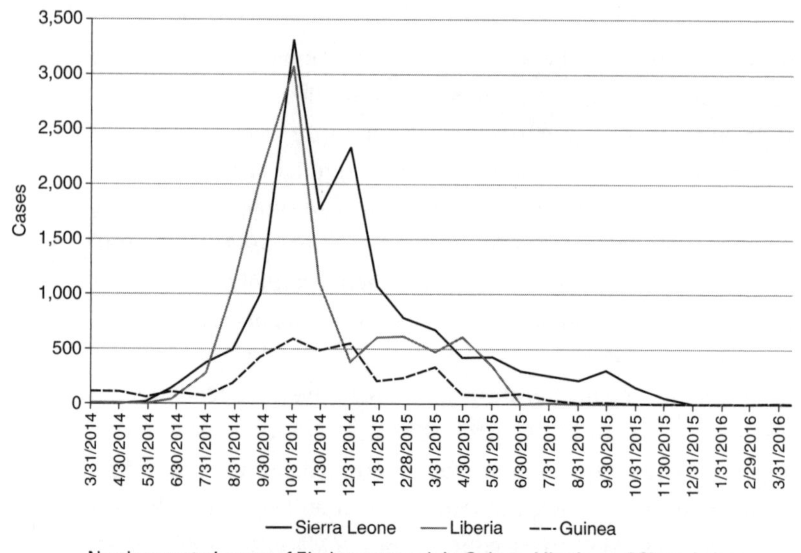

Newly reported cases of Ebola per month in Guinea, Liberia, and Sierra Leone, from March 2014 through February 2016. Numbers are a total of suspected, probable, and confirmed cases. Data from WHO Situation Reports (acquired from different sources before and after November 2014), as provided by the CDC.

The virus's rapid spread may also have resulted from a phenomenon known as superspreading, where a small number of infected people are responsible for generating the majority of secondary cases. These individuals—so-called superspreaders—are able to transmit the disease more easily than others for any number of reasons, including biological factors or proximity to others within the community. Superspreading has been a leading factor in previous infectious disease outbreaks, including SARS in 2003 and MERS in 2012.[20] Statistical models suggest that in Sierra Leone, just 3 percent of people with Ebola may have been responsible for an estimated 61 percent of cases.[21]

The progression of the disease is horrific. During the West African epidemic, Ebola patients spent on average six days in a hospital before dying.[22] The first symptoms to appear, about two weeks after infection, are indistinguishable from those of malaria or the flu. The longer the symptoms persist, the worse they become. It is commonly believed that Ebola patients die by bleeding out, but hemorrhaging was reported in less than a quarter of cases during the epidemic, and it is not the proximate cause of death in most Ebola patients. The majority of patients die from complications of severe dehydration due to fluid loss from vomiting and diarrhea. It is direct contact with these body fluids that poses the highest risk of transmission. Blood, sweat, and saliva late in infection are also considered contagions.

Not all who contract Ebola die. Some don't even get sick after being infected. While there is no known cure for the disease, the body's own immune system may be able to fight it off if symptoms are treated. Treatment includes intravenous fluids and electrolytes, blood pressure medication, and supportive care to replenish the massive fluid loss and allow the immune system to build up antibodies. The earlier a patient is given supportive care, the better the chances of survival. There may be other reasons why some people survive an infection while others die, such as age and underlying health. Some research suggests that genetics may play a role. The way some bodies respond to the infection—with either an aggressive, overactive immune response or a more targeted response—may affect the chance of survival.[23]

Even if an individual survives, the virus can remain hidden inside the body and cause complications or chronic ailments. The immune system remains weaker than it was before, as do the nervous system and cartilage. A 39-year-old British nurse who survived after contracting Ebola in Sierra Leone later developed life-threatening meningitis as a result of the virus lurking in her brain and spinal cord.[24]

The brain and central nervous system are among the places known as "sanctuary," or "immune-privileged" sites—pockets in the body that are protected from the body's immune response. Other sites include the eye and the male genital tract. There is evidence that the virus can replicate in the scrotum and remain in the body for up to five hundred days.[25] The first documented case of sexual transmission of Ebola to a woman with no risk factors other than having had sex with a survivor occurred during the 2014–2016 epidemic. Her partner's semen tested positive for the virus even though he no longer had symptoms. Even someone who survived Ebola, then, could still pass it on to others or even relapse themselves.

The incidence of Ebola was highest in adults between the ages of 35 and 54, probably because people in that age group served as caregivers and health workers.[26] Children, though they had the lowest incidence of disease, still faced significant risk because their caregivers were often exposed to the virus. In the Maforki Ebola treatment unit in Port Loko, Sierra Leone, children represented as many as one-third of Ebola patients at any given time.[27] Children typically show symptoms more quickly after infection, and the disease progresses more rapidly than in adults. The disease rate in children may actually have been higher than reported; many children were cared for at home, and others were not taken to a clinic because their caregivers were also ill.

The fact that the outbreak occurred in three separate countries also contributed to the spread of the disease. Where international responders saw national borders, many locals saw a borderless region within which they traveled freely.[28] Some children from Guinean families went to school in Liberia. Kenema, one of Sierra Leone's largest cities, is accessible to both Guinea's forest region and Sierra Leone's capital city, Freetown, which made it a key location for movement of the virus. Though citizens in the area travel easily across borders, the countries are governed separately, and international responders prepared separate response plans. "It was almost as if there were three ball fields and everyone was playing separate ball games, yet they were overlapping each other," said Michael Osterholm, director of the University of Minnesota's Center for Infectious Disease Research and Policy.[29]

The lack of coordination was particularly devastating for Sierra Leone. Hundreds of cases had appeared in neighboring Guinea and Liberia after March 2014 and had started quietly spilling over the border. By the time the virus was recognized in Sierra Leone in May, it was already pouring in like a tidal wave. Confirmed cases in Sierra Leone tripled as each week passed.

Sierra Leone ended up being the hardest hit of the three countries. Nearly 9,000 people were infected with Ebola during the epidemic, 2,000 more confirmed cases than in Guinea and Liberia combined. The virus was found in 114 of 150 chiefdoms in Sierra Leone, with the more densely populated chiefdoms having more confirmed cases.[30]

Many Sierra Leoneans likened the outbreak's devastation to that of the civil war—but this time, they were grappling with an invisible enemy. September 2014 saw a large uptick of cases across the region. An estimated 718 confirmed and probable cases and 289 deaths were reported in Sierra Leone, Guinea, and Liberia in the week of September 8–14 alone. The case reports are likely an underestimate, since some of the infected never made it to hospitals, and some clinics offered incomplete data. Scientists predicted that without drastic improvements in control measures, the number of cases across West Africa would increase from hundreds to thousands per week by the next month.[31]

Other than the scale of the outbreak, this occurrence of Ebola was similar to the ones that had occurred in the past in terms of the physical symptoms of the disease, the incubation period, the length of illness, the rate of infection and transmission, and the risk of death. "It is therefore unlikely that the particularly devastating course of this epidemic can be attributed to biologic characteristics of the virus," concluded Dr. Jeremy Farrar of the Wellcome Trust and Dr. Peter Piot of the London School of Hygiene and Tropical Medicine.[32]

Certainly, dysfunctional health systems and lack of trust in authorities in a war-ravaged region contributed to the devastation. But among the most destructive factors were international indifference and what we call outbreak culture.[33]

THE CRUCIBLE OF OUTBREAK RESPONSE

We forget the lessons we learn from outbreaks, so we make the same mistakes.

—Nahid Bhadelia, National Emerging Infectious Diseases Laboratory, Boston University

DR. SHEIK HUMARR KHAN'S death released a coil of panic for many Sierra Leoneans. Even those who had questioned the reality of the Ebola outbreak—suspecting that it was a hoax or a government conspiracy—began to believe in its existence after Khan died. Many wondered: If no one from the international community was willing to step up and help their beloved Ebola fighter, what chance would they, as ordinary citizens, have to survive? Like the virus itself, characterized by its long, branched, and sometimes spiraled filaments, the outbreak, too, had spiraled out of control.

The decision not to treat Khan with the experimental drug ZMapp remains swathed in controversy. As Khan lay dying, Nancy Writebol, an American nurse volunteering in Monrovia, Liberia, was diagnosed with Ebola, and the next day, so was her colleague

Kent Brantly, an American volunteer physician. The National Institutes of Health arranged for the vials of ZMapp that were still sitting in the freezer in Kailahun, near Khan's bedside, to be shipped to Monrovia. Nine days after Brantly showed the first symptoms of Ebola, he received one intravenous dose of ZMapp. Within hours, his breathing improved and the rash that covered his body disappeared. Writebol, who appeared to be far sicker than Brantly, was given two doses of ZMapp. Both consented to the treatment after having been told about the potential risks and benefits and informed that ZMapp had never been used on humans. Right after treatment, within days of each other, Brantly and Writebol were evacuated on a jet that was contracted by the US State Department. Both were treated at Emory University Hospital in Atlanta, Georgia—where Writebol received another dose of ZMapp—and both survived.[1] A Department of Health and Human Services independent panel that evaluated the Ebola response later concluded, "Differing perspectives on the most appropriate ways to use and evaluate investigational vaccines and treatments contributed to incomplete evaluation of the efficacy of these products."[2] Questions swirled among residents of the Ebola-stricken region as to why white Americans were airlifted to safe zones, offered exceptional care, and administered experimental drugs, while Africans were left on the streets and in overcrowded hospital wards to die.

All who were involved in Khan's care acknowledge the difficulty in deciding what to do, but those closest to him could not reconcile themselves to the fact that they had passed up a chance, however uncertain, to save him. "It was a mistake for so many reasons," Tulane University immunologist Robert Garry remarked. "But I don't think it was a value judgment of Americans over Africans."[3]

Khan's situation placed decision makers in an exceptionally difficult position. There were other Ebola patients in the ETU who were equally sick and were not being considered for special care.

But the argument could be made that Khan was revered by Sierra Leoneans, especially in Kenema. Perhaps he should have been considered a special case because he risked his life to save others from the virus that had infected him. Everyone who knew Khan believed that as an expert on Ebola, he could have made an informed decision to try the experimental drug. Still, ZMapp had never been used before in humans, and the policy of MSF, which was the presiding agency in Kailahun, did not allow for its use even during an emergency. One WHO clinical responder who visited Khan in Kailahun suggested personally administering ZMapp to Khan and supervising his subsequent care. MSF responders denied the request in large part due to ethical concerns of singling out any one patient.[4]

What about evacuation? Many governments were hesitant to accept any foreign national infected with a highly contagious disease into their country.[5] Some argued that evacuating Khan might set a precedent for more evacuations. Then the same question would come up for the next Ebola patient: "Why him and not me?" During an outbreak, who decides who gets to leave and who must stay? Where do government health agencies draw the line? How much do we risk to save one human life? No one had answers to these complex questions.

Sierra Leoneans who knew Khan, and even those who never met him but heard his story, believe there is nothing to debate. They believe more should have been done to save him. The perception that race determined survival in this circumstance furthered fear and mistrust among many West Africans, who were already skeptical of large agencies and even of their own government. "They murdered Dr. Khan," said Simbirie Jalloh, program coordinator of the Lassa Fever Research Program at Kenema, in despair. "Even if they didn't take him out of the country, even if they [just] took him out of Kenema [and kept him in Kailahun], he should have survived."[6]

On August 11, 2014—less than a month after Khan's death—a group of experts convened by WHO reached a consensus that the use of experimental medicines and vaccines under the exceptional circumstances of the Ebola epidemic is ethically acceptable.[7] For Khan, the decision came too late.

Khan's work and his death are at the heart of a story that is about more than a single hospital in one African town, or the deliberation about how to care for one of Sierra Leone's most prominent doctors. Every facet of his work, his disease, and the minutes leading up to his death testify to the fractured system of outbreak response. Forces much larger than Khan and his staff made it impossible to launch a timely, effective, transparent, and collaborative global response.

Outbreaks are a crucible, a melding of elements brought together in an intense pressure-filled environment. The environment is volatile, the situation is often unclear and rapidly evolving, and the pathogen is life-threatening. The scenario provides a test of the human response to unpredictability.

"People should never forget how tense, inevitably chaotic, and frightening these situations are," said Jeremy Farrar, director of the London-based Wellcome Trust, in an interview. "When you're in the middle of these situations, on a personal, professional, and organizational level, nobody can underestimate how frightening that is." The "secondary issue," as Farrar called it, is the politics-driven chaos that unfolds during a large-scale outbreak.[8]

During the crisis, a culture can form that is as lethal as the pathogen itself. Outbreak culture, as we call it, develops among individuals and institutions when the worst human instincts and behaviors are released by the crucible of a quickly spreading disease. Multiple factors drive its creation, from political motivation or desire for personal gain, to overwhelming fear and isolation. The resulting behaviors can lead to confusion and misconduct, adding to an already chaotic environment.

Many stakeholders and decision makers sit thousands of miles away from where the infected are dying. Clinical responders, scientists, and staff with government and nonprofit agencies and for-profit companies of every size and scope are shackled by internal politics and policies that hold the keys to improving outbreak response. The Khans of an outbreak become collateral damage. After the epidemic dies down, another emerges, possibly with a different name in a different part of the world, and the sequence plays out again.

From the outside, these problems are often invisible, and outbreak response appears focused and organized: governments and nongovernmental agencies are sending money, supplies, and personnel, and academics are studying samples. What is known to responders behind public view, however, is that outbreak response is often filled with perverse incentives, nefarious practices, and blame. This same culture has been identified in many large-scale disease outbreaks preceding Ebola.

Several common features characterize the divisive environment that develops during a serious infectious disease outbreak. Rivalry among high-profile agencies to secure their place in the narrative of meaningful response. Competition among researchers to be the first to detect, collect, and analyze pathogen samples at all costs. Reliance on perception, not science, for policymakers' decisions. Poor communication and coordination. Stalling among large agencies for political reasons. Frustration among clinical workers caught in the middle between the decision makers and the dying. Misunderstanding of the local culture. Pressure to blame someone or something. Fear within the local people, who are unaware of various stakeholders' compromised positions. Promises to "do better" next time.

All of these features have little to do with the biological makeup of the pathogen or the disease itself. In fact, a majority of outbreak responders whom we interviewed or who responded to our survey reported that they felt far greater stress from the political battles that went on during the outbreak than from the virus itself. Disease detection and response is a complex process with one inescapable truth: the culture that develops during a disease outbreak can be secretive, competitive, and at times, contribute to the spread of the disease, not its containment.

Clearly the pathogen plays a critical role in an outbreak, as it is the root cause. The nature of the pathogen embedded in a region, especially one with a fragile health system, should remain the primary concern, according to Jeremy Farrar. "If we lose sight that this was an outbreak of an infectious disease and we blame the epidemic on something existential, we'll be doing a great disservice," he said.[9]

Regardless of the pathogen involved or the location of the outbreak, the pattern of how the international community responds remains the same. According to Sheila Davis, chief nursing officer and head of Ebola response at the global health nonprofit Partners in Health, the knowledge that outbreak scenarios in general unfold quickly, especially in locations with limited resources, is enough to warrant swift action, build up capacity, and impose universal directives. There is no blanket protocol within the international community on responding to disease outbreaks—there's neither a mandate to implement certain protocols nor any accountability for following protocol. While postmortems at large global health organizations aim to draw lessons so as to prevent similar disasters from occurring again, the preventive actions taken often involve only individual organizations or groups and have not yet led to collective systemic changes. "We have been saying for 25 years that we need to get our act together," said Robert Garry. "But really we're waiting for people to die."

Rapid and effective disease outbreak response is a complex, multistep process requiring recognition, reporting, verification, and response.[10] In an ideal world, a unified communication system among public health agencies and strong local health systems in vulnerable regions would enable outbreaks to be swiftly detected and contained. Unfortunately, this is rarely what happens. Nearly every major infectious disease outbreak has faced challenges in detection and response, but the level of dysfunction at times during the 2014–2016 Ebola outbreak was exceptional. With Ebola, nearly six months passed from the time the first case was detected in West Africa to when international agencies like the WHO recognized and launched a large-scale response. By that point, there was an epidemic killing thousands of people across the region and threatening to spread to other continents.

As has occurred during many public health emergencies, thousands of people volunteered with nonprofit and nongovernmental organizations to respond to the Ebola outbreak. Many responders risk their lives during infectious disease outbreaks, and, in the case of Ebola, hundreds died trying to save others and limit its spread. Our survey found that many who volunteered felt their capabilities were restricted and their efforts undermined by a destructive culture. More troubling, manipulation and bad intentions were often cloaked in justifications of the greater good.

Despite the serious complications, many who are deeply involved in outbreak response did not find the situation surprising, given the intense environment during outbreaks. "Everything happened so fast with Ebola, there is no way that we should think everything should have gone smoothly," said Davis.[11] Still, the longer the time between epidemics, the more likely we are to forget the lessons learned.

Medical historian Charles Rosenberg describes an epidemic as a "dramaturgic event" with three acts: a "progressive revelation," in which communities gradually accept the presence and spread of a contagious disease, followed by "managing randomness," when people try to identify a cause and often blame others, and finally "negotiating public response," during which measures are taken to control the epidemic.[12]

A prominent feature of the first act, "progressive revelation," is that recognition is often delayed. This delay may be caused by the very government agencies that are supposed to track disease outbreaks. By regulation, countries—often through their public health agencies—are required to report outbreaks of certain diseases to the WHO.[13] But newly emerging diseases may be overlooked because of inadequate healthcare infrastructure, including lack of standard surveillance systems and rapid diagnostic testing. Even if a disease is detected, governments may be reluctant to publicly acknowledge a potential epidemic for fear of disrupting public order and drawing negative international attention. The economic risk of reporting a disease outbreak—such as bans on trade and travel—is enough for governments to defer acknowledging a public health crisis. Attempts by government agencies to prevent disease surveillance and reporting have contributed to the spread of otherwise containable diseases, in both humans and animals. In October 2011, for example, federal government officials in British Columbia reportedly intimidated and tried to undermine scientists who detected a contagious viral disease in farmed salmon.[14]

Human infections are much more difficult to conceal, and doing so can have dire consequences. Although the first cases of a new disease, later dubbed SARS (severe acute respiratory syndrome), appeared in China in November 2002, the Chinese government did not officially report cases to the WHO until February 2003. Within months, the disease had spread beyond China into 37 countries. Almost ten thousand people were infected, and nearly a thousand died.[15] The Chinese government did not publicly acknowledge the epidemic until March 2003, when China's health minister told the public on state-owned television, "The illness has gradually been brought under effective control."[16] But the virus was still spreading and killing people, and a month later, the health minister was fired.

Delays in reporting are not always intentional; sometimes the problem has to do with inadequate surveillance and reporting systems. In October 2005, a cluster of cases in Sudan of a disease with symptoms similar to smallpox was reported by Médecins sans Frontières (MSF) five weeks after the first case of the disease emerged. The disease was later identified as monkeypox, a disease that is not very easily transmissible. Had the disease been more contagious, the five-week delay in reporting could have led to a large outbreak and a public health emergency.[17]

Once a disease outbreak has been recognized, we move on to Rosenberg's second act: managing randomness by blaming or scapegoating the victims. In the early twentieth century, many Americans blamed Irish immigrants for bringing typhus, tuberculosis, and cholera into the country. Instead of working to help them, some doctors accused them of being "exceedingly dirty." Immigrants were refused medical care, and many "wandered starved and half naked along the Canadian border."[18] Refusing to help people, we have since learned, does not stop an outbreak.

The underlying impetus for blame among the public is often fear. Hollywood movies have been particularly effective at stoking that fear. *Andromeda Strain* (1971) features a team of scientists battling a pathogen of extraterrestrial origin, brought to earth by an outer-space mission. *Outbreak* (1995) focuses on a fictional Ebola-like virus called Motaba, which originates in Zaire and later emerges in small-town America. These and many other movies about epidemics—a long list that includes *Contagion, World War Z,* and *28 Days Later*—all follow the same plotline: discovery, chaos, death, survival of a lucky few, and heroic scientists who come to the rescue. These films play into the audience's emotions. It is this sense of fear and uncertainty, a very human response, that allows emotions to override science during outbreak response.

Fear and blame are also frequently fanned by the media. In the United Kingdom, for example, media coverage of the SARS outbreak perpetuated the stigmatization of Asians and assigned them responsibility for the epidemic.[19]

Scapegoating of particular groups was especially widespread during the AIDS epidemic. AIDS—which at the time was not yet named—was first reported in 1981 in five gay men in Los Angeles. By the end of the year, there were 270 reported cases, and 121 of those individuals had died.[20] In the years following, as scientists, politicians, and the public sought to make sense of the epidemic, people began to call AIDS a "gay plague." Pat Buchanan, speechwriter for President Ronald Reagan, wrote that homosexuals have "declared war on nature, and now nature is exacting an awful retribution."[21]

AIDS was also detected among intravenous drug users, and, in July 1982, the CDC announced that 34 Haitian residents in the United States had infections that were similar to those that had been afflicting gay men and drug users.[22] In 1983, the CDC described the major routes of transmission of the human immunodeficiency virus (HIV) that causes AIDS and identified the four risk groups for the disease, which became known as the "4-H" club: homosexuals, hemophiliacs, heroin users, and Haitians. Nurses like Sheila Davis, who cared for AIDS patients in the 1980s, often faced public shaming for interacting with groups that had become outcasts. "With HIV, so much of the disease was our reaction to it," said Davis. "How we worked with it was framed by stigma."[23]

When the fifth "H" emerged—heterosexuals of any gender, race, or socioeconomic background—it became harder to know where to place the blame. The July 1985 edition of *Life* magazine featured photographs of a young woman, a young heterosexual family, and a saluting soldier. The cover article detailed the lives of three individuals who were hardworking Americans. They were patriotic. They were not promiscuous, nor were they drug addicts. They were not the face of AIDS that had been etched through years of stigma, yet they were infected, through blood transfusions or long-term partners. Bold red letters stamped across half of the magazine's cover read, "Now No One Is Safe From AIDS." And indeed there was some truth to this appeal to fear: Viruses know no boundaries.

Pathogens don't distinguish between nationality, religion, or political beliefs. They also don't respect country borders. Yet during the third act of an outbreak, "negotiating public response," when attempts are made to control the pathogen's spread, politicians and policymakers tend to ignore this fact. In every major outbreak, many US policymakers have advocated for stricter border controls or for stopping immigration from countries where the disease is found. During the AIDS epidemic, the US government denied people who were HIV-positive entry into the country.[24] In 1992, the International AIDS Society relocated its annual conference, originally scheduled to take place in Boston, in protest of countries that restricted short-term entry of people living with HIV. The conference—ironically themed "A World United against AIDS"—was moved to Amsterdam.

It's often politicians who advocate for border control, not scientists—and there's a reason for that. "Protecting" the border plays into the public perception that pathogens can be stopped by a wall. But restricting movement in and out of a country restricts the flow of necessary aid to combat an epidemic and prevent further spread of the disease. Science tells us that the best way to prevent and control epidemics is to enter a region in order to build up health systems, conduct research during and between epidemics, care for the afflicted, and swiftly send in personnel and supplies.

The response to Ebola was slow in part because agencies were reacting to accusations that they had based their responses to previous epidemics on public fear rather than actual risk. The WHO, which had been criticized for overreacting during the 2003 SARS outbreak and the 2009 swine flu pandemic, hesitated to declare Ebola a public health emergency and did not begin outbreak response measures, despite repeated calls to action from other responding organizations. The difference in the case of Ebola, however, is that people on the ground in West Africa knew that a major epidemic was brewing.

The response to outbreaks over the course of history shows a pattern of discord between flow of information and action. Indeed, with recent major outbreaks such as swine flu in 2009, Ebola in 2014, and Zika in 2015, each of which was declared a public health emergency by the WHO, the delay may have had less to do with inadequate surveillance systems than with an inability to mobilize collective action. In a study of the causes of delay in response in these three outbreaks, public health researchers Steven Hoffman and Sarah Silverberg calculated that poor mobilization was responsible for 1.9 times more delay than insufficient surveillance capacity. The quickness with which a public health emergency was declared did not depend on the severity of the outbreak. Ebola had the highest fatality rate of the three, yet it yielded the slowest response. Nor did it depend on the number of countries affected: swine flu had only spread to three countries when it was declared a public health emergency, while Zika had spread to 21. Nor did a quicker response hinge on the number of people at risk for infection. Hoffman and Silverberg concluded that the speed of response was best explained by two factors: the ease with which a disease is transmitted, and the likelihood that it might affect people in the United States.[25]

The time between reporting cases and getting boots on the ground is perhaps the most critical for saving lives. This period is also one of moral reckoning. Agencies and countries know there is something wrong and that they could be doing something about it, but they hesitate. In some cases, poor coordination between governments, UN agencies, and nongovernmental organizations compromises timely emergency response.[26] There is also a certain amount of jockeying for control among political leaders. Governments of affected countries play host to a number of foreign agencies, clinicians, companies, and scientists, whom they permit to enter during the outbreak. Each entity is working in various capacities, but there are no clear guidelines indicating who has jurisdiction over the cases handled, the data collected, or the research conducted. Thus follows a politically motivated reach for control by all, often in the form of bogus policymaking, data hoarding, and restrictions on research, and in the worst case, direct attempts to obstruct others' progress.

In 2006, Indonesia's minister of health claimed "viral sovereignty" over samples of avian flu virus that had been collected from patients in the country. This concept asserts that viruses are the sovereign property of the individual nation in which the samples were taken—even though they cross borders and could pose a global threat. Many avian flu outbreaks had occurred in Indonesia in both poultry and humans, and dozens of strains of the virus had been identified. But Indonesia had shared with the WHO only two samples from people who had been infected, and the country had also stopped notifying the WHO of outbreaks.[27] In this case, the Indonesian government believed it was protecting itself both from foreign researchers, who might profit from treatments or vaccines developed from the virus samples without making those products available in Indonesia itself, and from governments that might use the data to build their biological warfare capacity.

Indonesia's actions are an example of the sense of ownership that can develop over a contagion or the region where the contagion emerged. We believe that research and data dissemination should be guided by the principle that pathogens do not belong to the country where the outbreak takes place, or to any one group or person studying them. Government hoarding of samples or data is an especially dangerous ideology for developing countries that do not have the resources to control the pathogen. Still, the belief persists mainly because of continued scientific exploitation of regions overcome by disease. Dr. Peter Piot, director of the London School of Hygiene and Tropical Medicine, describes this practice as the "cowboy factor."[28] Put simply, scientists "parachute" into hot zones, stay long enough to gather samples, conduct a few tests, and get out. Researchers use the data to analyze the pathogen and develop everything from standards of care to new vaccines and drugs, but the host country might not receive the benefits of any of these products for years, if at all.

Patenting viruses, for example, can lead to restricted access to data and even efforts to delay others from developing competing products. This scenario played out publicly in 2013, when Dutch scientists who patented the coronavirus that causes MERS (Middle East respiratory syndrome) were accused of hoarding the virus samples, potentially keeping other scientists from developing diagnostic tests.[29]

Poor communication to the public is another feature of outbreak culture that can influence the course of outbreak response. Often the psychological effects on people who fear what they perceive as an unpredictable and uncontrollable event have a larger impact on the severity of the outbreak than the physiological effects of the pathogen. Fear is a powerful motivator, and it can lead to behaviors that have major sociopolitical and economic cost. The SARS epidemic, for example, although it killed nearly a thousand people, produced economic effects that were far out of proportion to the health effects.[30] During the height of the outbreak, there was an 80 percent reduction in air traffic and a 50 percent reduction in retail sales in Hong Kong.[31] Much of the fear was unjustified. The virus spread relatively slowly, and a person was only infectious when symptomatic, meaning that classic outbreak containment measures such as isolating individual cases and tracing contacts proved to be effective. Many of the protocols followed during the outbreak had more to do with perceived risk than with actual risk.

While it is up to public health officials and governments to make decisions based on actual risk, these decisions are often driven by emotional, political, or economic demands of the public. In China, visible public health activities such as wearing masks in public and conducting infrared screenings at airports probably did little to curb the spread of the disease, but they provided more public reassurance than containment measures that actually worked, which may have helped reduce the economic impact.[32] Despite the usefulness of such measures at reassuring people, it is worth questioning the intentions of agencies, organizations, and individuals and the potential to exploit public fears for political gain.

Highly visible organizations and individuals responsible for outbreak response feel incredible pressure to impose decisive and noticeable measures; they need to build the public's confidence by showing that they are doing something of perceived benefit, even if the actions have no scientific merit. This was the case during the Ebola outbreak, when some politicians called for travel bans and closing their borders to West Africans and healthcare providers returning from the affected region. Some individuals traveling to the United States were ordered to be quarantined, even though they had had no direct contact with infected individuals. These directives seemed intended to quell public fears, yet they may also have been motivated by political considerations. From a scientific perspective, these actions would have made no difference in the outbreak's trajectory.

Even when organizations enter volatile environments with good intentions, they may exacerbate a crisis because of the political and anthropological landscape of the region. For many people, perceived risk has less to do with the chances of contracting the disease than with the fear of strangers entering the community and disrupting a way of life. Such was the case during the 2010 cholera outbreak in Haiti, which killed at least ten thousand people and sickened hundreds of thousands more. Some parts of Haiti, a country whose people have been marked by poverty, colonialism, and slavery, violently resisted foreign responders.[33] As it turns out, however, the UN, the agency tasked with containing the outbreak, was most likely the culprit behind its onset in the first place. Investigations have shown that the epidemic originated from sewage released into a river by UN peacekeepers from Nepal, which was undergoing a cholera outbreak at the time.[34] This disaster caused by foreign intervention, though unintended, served to confirm local perceptions. Although the UN has since apologized for its role, distrust of foreign responders and the demand for accountability and compensation continue to this day.

Resistance to response efforts, which also occurred in parts of West Africa during the Ebola outbreak, result from focusing solely on containing a pathogen, and underestimating how important it is to understand the anthropology and history of a region. We discuss this aspect of outbreak response in more detail in Chapter 5.

The media has an important influence on public perceptions of risk, since the general public gains knowledge about outbreaks primarily through media outlets. The pressure placed on these outlets to disseminate information in real time can lead to conflicting reports during an unfolding situation. Especially during an outbreak of a pathogen about which there is not much scientific knowledge, like SARS, information presented publicly may be based mostly on "opinion, guesswork, and preliminary results." The reciprocal relationship of the public feeding the demand for media coverage and media outlets feeding into the frenzy leads to what researchers have concluded, in the case of SARS, was "excessive, sometimes inaccurate, and sensationalist" media coverage.[35] The same pattern characterized other disease outbreaks both before and after.

Scientists are often urged to inform the public of outbreaks as soon as they are detected. However, in cases like Ebola, scientists are learning about the virus and its origins even as it is spreading. While the ability to conduct research and share data in real time is paramount to improving preparedness and early response, there is no mandate for any individuals, groups, or countries to share data or other information pertinent to the pathogen or its spread, especially while the data are being collected.

A lesser known but equally destructive aspect of outbreak response is deliberate venality by private companies, scientists, and other stakeholders who seek to exploit an epidemic for personal gain. Outbreak culture can be one of corruption, competition, and sabotage, even at an individual level.

Scientists are sent to an outbreak "hot zone" to collect samples from infected people and other data to take back to their labs to study. The goal is to better understand the nature of the pathogen and origin of the epidemic, and, in the case of Ebola, develop a diagnostic test to detect the virus and determine the best approaches for prevention and treatment. These efforts are critical for controlling an epidemic. As in a war zone, however, an outbreak zone can bring out the worst in people.

"All these people are vying for the same piece of data," said Stephen Gire, chief scientific officer at the biomedical startup NextGen Jane. "It can be difficult to get work done to begin with, but they also have their own self-interest."[36] Gire had been working with the Sabeti Lab and traveled to Kenema twice during the Ebola outbreak. Before that, he set up a lab in the Democratic Republic of the Congo to study monkeypox. He was so inspired by his work in Congo that he got a tattoo of a monkeypox particle and started the NGO Congo Medical Relief to provide medical supplies and equipment to the region. Gire was instrumental in collecting and bringing back Ebola samples that were then tested, sequenced, and analyzed to provide the first comprehensive look at the strain of Ebola that had evolved during the outbreak. As we detail in the next chapter, however, the toxic environment during the outbreak response created serious obstacles for his work. This experience caused him to leave the Sabeti Lab in February 2015.

Division among research groups has occurred in numerous other large-scale epidemics, but Peter Piot told us that it was not always the case. "In epidemics, in crises, the best and worst behavior comes out in people," Piot said. "People risk their lives, but it's also the hope for glory and discovery." Piot, who was part of the team that first discovered the Ebola virus in 1976, noted that the three research papers publishing data from samples acquired during that first known outbreak were prepared as a coordinated effort between research groups. The researchers also agreed to publish a single epidemiological brief in a bulletin published by the WHO. "It was a smaller outbreak and very few people were involved in it," Piot said, adding that there was also very little public interest in the outbreak. He admitted to a certain amount of naivety in the research and publication process, since his involvement in the discovery occurred very early in his career. "I thought that [process] was 'normal,'" Piot said. "That obviously did not happen in West Africa," he added, referring to the lack of cooperation during the 2014–2016 epidemic.[37]

This kind of cooperation also did not happen during the first decade of the AIDS epidemic in the 1980s. "There was, of course, the fight over who 'discovered' HIV," recalled Piot, who was one of the first scientists to study AIDS in Africa. "That was probably the most public and visible aspect of what was going on in the AIDS field." Following a bitter dispute over who should receive credit for discovery of the HIV virus, the Nobel Prize in medicine was eventually awarded to a pair of French researchers, Luc Montagnier and Françoise Barré-Sinoussi. Later, an out-of-court agreement stipulated that American scientist Robert Gallo be recognized as a co-discoverer of the virus and share patent royalties.[38]

As a result of this clash, the National Institutes of Health, the US government's principal agency for biomedical research support, implemented sweeping changes in the way it reviews research proposals for funding. The changes included implementing standard protocols for gathering data, and incentives for sharing data with others.[39]

"Behavior change only came when there was a stick and a carrot, and [the latter] was big NIH funding," Piot said. Still, it wasn't enough to halt competitive behavior, especially among large agencies. "What shocked me the most in the West Africa outbreak was that public institutions like CDC or WHO, those paid by public money, were not sharing the information, not sharing the data, not sharing the samples."

Large institutions were perceived as monopolizing access to data, which did not allow other teams to independently analyze the samples and provide information to people working on the ground. "Contemporary molecular techniques could offer precious information as long as it's there on time," Piot said.

Access to samples would have helped responders understand transmission chains and gain insight into why the outbreak died down at one point and then seemed to reignite overnight. "When it comes to public institutions, one of the basic principles of good governance in a democracy is that information is owned by the public," Piot pointed out. Data hoarding, he said, "was a gross violation of good behavior."

Pressure mounts among researchers to be the first to discover a pathogen or an outbreak. Public health agencies strive for publicity in their effort to maintain legitimacy. For-profit companies aim to create and license technologies essential to outbreak response. It might appear that aggressive, highly motivated people and rushed work are exactly what is needed to combat a fast-spreading virus. But aggression and urgency can lead to deception, secrecy, and attempts to put obstacles in the way of competitors. "Human nature is that we are invested in our own gains and our own philosophy and our own organizations," said Sheila Davis of Partners in Health.

Although these impulses are critical to pushing forward research and development, the incentives are not set up to benefit those who are in greatest need. In an article in *Nature* on how phenomena in developing countries are studied for the benefit of foreign institutions, Dr. Brian Conton of Freetown, Sierra Leone, wrote that local scientists are often not given authorship in scientific publications. "And worse, research can become skewed to fit the demands of Western academic careers, rather than solving the problems that the disease causes where it occurs."[40]

Many of those who are involved in outbreak response recognize this practice. Conton refers to it as "helicopter research." Dr. Joseph Fair, who served in an advisory role to Sierra Leone's Ministry of Health and Sanitation during the outbreak, calls it "scientific colonialism."[41] In an interview, Fair noted that some responders travel to the site of an outbreak under the auspices of capacity building, but instead they work for their own personal and professional gain, exploiting the environment, people, and resources directly affected by the disease.

Fair first arrived in Sierra Leone in March 2014. He left in May when cases decreased, but then returned in July when they began to spike. Initially working in a volunteer capacity, Fair began to receive pay for his advising role in August through a contract from the US Department of Defense. Throughout the epidemic, Fair was among those petitioning international groups to send staff and supplies. Between September and December, he spent much of his time traveling between Liberia and the Ebola emergency operations center in Sierra Leone. He briefly returned to the United States in September to update members of Congress on the situation alongside then CDC director Dr. Thomas Friedan, among others.[42]

"This outbreak [Ebola] was like researchers gone wild," said Fair. Now a senior adviser with the nonprofit research organization MRIGlobal, Fair was one of the founders of the biotechnology company Metabiota, which was operating in Kenema Hospital at the start of the outbreak. Speaking generally about what he observed while in his advising capacity with the ministry, he said, "I have never seen people drop their ethical standards so fast."

Among other things, Fair witnessed researchers who were cleared to conduct unrelated research switch gears to work on Ebola, apparently without clearance. Some scientists broke the rules under the guise of conducting research for a "good cause," but in some instances, they may have crossed an ethical line, he said.

Much of what occurs during an outbreak is not in the public eye, but during the Ebola outbreak, some glimpses emerged of the underlying chaos. Feuding between responders—especially at the local level—moved from behind the curtain to center stage, and the public's perception of the inadequacy of experts heightened fears. "This happens a lot more than we see," said Sheila Davis.

"People are not focusing on the effort to help but on the political side of things. It forces people to take sides."

Early in the epidemic, feuds put responders who represented companies, organizations, and agencies on the defensive and prevented them from cooperating and listening to each other. When aid finally arrived, the lack of coordination and collaboration led to wasted efforts and money. The US government designated over $5 billion to combat the outbreak domestically and internationally. The response, however, has been critiqued for building Ebola treatment units that were never staffed and received no patients, operating dangerous clinical trials, and overreacting to unfounded public panic.[43]

Some companies followed the money and reportedly sought to profit from the frenzy. The CBS news magazine show *60 Minutes* aired a story in 2016 looking into claims that a medical equipment manufacturing company, Halyard Health, had knowingly provided defective protective gowns that were used by Ebola responders in the United States and abroad during the height of the outbreak. The gowns, it was claimed, did not meet standards of impermeability and could allow fluids to leak through. At the time, Halyard Health denied that it had put any healthcare workers at risk.[44] In 2017, a Los Angeles jury reached a $454 million fraud verdict against the company and its former parent company, Kimberly-Clark, for the marketing and sale of surgical gowns that were known to be defective.[45]

Foreign responders are not the only ones who exploit outbreak culture. Governments of the three affected countries in West Africa also took advantage of the delayed response and poor coordination among outside agencies. As we discuss in Chapter 7, a great deal of financial fraud went on in West Africa during the epidemic.

Many of those who responded to the Ebola outbreak also reported witnessing threats and intimidation of some kind. One survey participant wrote that international organizations were met with "threats from government to stop our Ebola response if resources were not channeled through them." Many NGOs threatened to reduce or stop aid to countries in response to government demands. Political groups and officials put pressure on government and international agencies to use specific suppliers. Groups often wielded their political influence to avoid following Ebola protocols as well. Chapter 3 further discusses what our survey results uncovered about the prevalence of unethical and illegal practices during the Ebola outbreak.

Many of the worst features of outbreak culture—from general discord in response efforts to flagrant systemic corruption—resulted from the lack of a central governing system to create, implement, and oversee standards of international response. Without central governance of public health emergencies and a regulatory body to ensure adherence and accountability, standards of behavior are determined by the organizations themselves. Organizational justice theory posits that in organizations that are perceived to be fair, transparent, and just, workers tend to be loyal, trustworthy, and committed. By contrast, if organizations are perceived to be behaving in an unfair, arbitrary, or biased manner, workers will show a lack of commitment, trust, and loyalty, and they may misbehave or perform poorly. A worker who sees others behaving unethically without any repercussions will be more likely to engage in the same behavior—it becomes part of the culture.

There is no international governing playbook for what to do during a serious disease outbreak, and guidelines cannot be created in the midst of a crisis. A key part of forging any directive, according to Sheila Davis, is the commitment to integrate the perspectives of local people at an early stage, recognizing their invaluable perspective and expertise. "If we said now, 'what happened with Ebola? [What] could we have done differently?' [One response would be that] we did not take into consideration or ask those closest to the issue what the issues are that can lead to [the epidemic]," said Davis. "Responses and efforts are often decided far away from where the problems are. People don't really give enough respect to those who are most affected to ask, 'What can we do?'"

Every aspect of the pathology of outbreak culture that seasoned outbreak responders were already familiar with proliferated during the Ebola outbreak response. If it is any indication of how future outbreaks of rapidly spreading pathogens will proceed, the global community should be concerned. However, as we reckon with what went wrong, we must resist the temptation to assign blame. One of the respondents to our survey described the worst experience during the outbreak as "sitting in a government meeting and hearing everyone accusing everyone else for the disaster. The fear in the room was palpable." As Davis points out, "Our culture of blaming makes people not necessarily want to take responsibility for doing wrong."

The way in which the global community views outbreaks must change, because the current emphasis is on responding to, rather than preparing for, outbreaks. As we discuss in detail in Chapter 9, the interepidemic period—when no urgent response is needed—is critical to building up strong systems to prevent and prepare for an outbreak. This includes strengthening global systems focused on interagency coordination. Then, when an outbreak occurs, the focus should be on immediate resources for the affected area. For Ebola, that would have meant adequate funding and shipments of medical supplies, PPE, and personnel. Onsite work should be approached with compassion for, and interaction and involvement with, those affected. Building trust between authorities and local people will help the public hold steadfast to facts, not fear. Responding organizations should focus on accountability—besides ensuring that the pathogen is contained. Sheik Humarr Khan and hundreds of other West Africans died in part because there was no mechanism in place to hold responders accountable for an early, appropriate, and effective response. "The time of crisis is not the time you should be organizing these things," Davis pointed out.

These steps to building capacity, coordination, and accountability begin between outbreaks. Ebola always seemed to be one step ahead of the response, but things could have gone differently. The virus had flared up several times over the decades, and each time it had been quickly contained. This time was different. The virus seemed to feed off our outbreak culture.

THE CASE FOR COLLABORATION

The politics were worse than the virus.
—Sentiment expressed by several responders to the Ebola outbreak

THE LAST TIME MEMBERS of Pardis Sabeti's research team saw Dr. Khan in person was in July 2014. Sabeti Lab scientists Kristian Andersen and Stephen Gire had helped establish diagnostics for Ebola at the Kenema Government Hospital laboratory in March 2014, in partnership with Khan and Augustine Goba, head of the Kenema diagnostic laboratory. In response to the growing number of cases at the hospital, Gire returned in July with Nathan Yozwiak to aid in the response, help the hospital and its partners streamline diagnostic testing, and share critical information they were generating on the Ebola virus through genome sequencing—which could be used for improved diagnostics, therapies, and surveillance efforts.

When they arrived at the lab on July 3, they saw a very different environment than they had ever experienced at Kenema. The

(*Left to right*): Simbirie Jalloh, Mambu Momoh, Sheik Humarr Khan, and Augustine Goba of the Kenema Government Hospital stand with Kristian Andersen and Stephen Gire, who arrived from Boston with supplies to detect Ebola virus in April 2014 as the outbreak was declared in West Africa. *Courtesy of Stephen Gire/© Sabeti Lab.*

hospital was filled with tension and mistrust. Rumors were swirling. The influx of international agency medical officers, while initially welcomed, had only exacerbated this situation as new visitors unwittingly spread misinformation and confusion. "The virus was just a backdrop to an immense political strain that was breaking everyone," Yozwiak recalled.[1]

On the third day of his visit, Gire headed to the lab in the morning. The Lassa laboratory—as it was still being called then—was a single-story building located on hospital grounds, but separate from the clinical facilities. It was divided into a general clinical area for routine diagnostics, and a separate area for handling samples from suspected Lassa fever cases. Because of the recent surge in Ebola cases, the lab workers had shifted their focus from

Lassa fever to Ebola. A handful of local technicians were responsible for caring for patients, collecting and testing samples, and entering data and managing reports—a workload that became overwhelming during the outbreak. Gire and other researchers with the Sabeti Lab and Tulane University had been helping with the data entry and other tasks to keep the workflow going. They were also trying to streamline processes as the number of cases grew. That morning, however, as Gire approached the lab, he was met at the door by an unfamiliar medical officer from the WHO.

"You can't come in," she told him, standing in the doorway of the lab.

"Why?" he asked. He looked into the room behind her, which housed the machines and computers used for generating and storing data. The officer was the only one in the building. Other staff members had all gone out to take care of patients.

"We can't let you come in here unsupervised," she said.

Although Gire and Yozwiak's genome sequencing work—which involved storing and shipping excess, discarded clinical samples—had been approved by the Sierra Leone Ministry of Health and Sanitation (MoHS), WHO officials feared that their presence would hamper response efforts. The WHO was concerned that by conducting any research-related activities at all in the midst of the outbreak, researchers were taking up valuable lab space that could be used for more pressing tasks, such as immediate laboratory testing or resources for patient care. They also suspected that under the guise of helping with data entry, the researchers were taking data for their own use that they had not been authorized to access.

"I have to do my work, and this is where I work," Gire said, as he came in the door.

Although the MoHS was the ultimate overseer of the hospital and signed off on all who were there, no ministry staff members were present in Kenema. Instead, the WHO regulated sample testing, analysis, interpretation, and reporting of results to the ministry. Many who worked at the hospital reported that the WHO had become a filter for all operations on hospital grounds.

The San Francisco–based epidemic tracking company Metabiota, which had shared lab space with Tulane since 2009, had been selected by the MoHS in March 2014 to monitor suspected cases of Ebola in Sierra Leone and to aid in response efforts after the virus had been found to be spreading in Guinea. Emails obtained by the Associated Press found that the WHO had accused Metabiota of attempting to "systematically bypass and marginalize [the] WHO role" by drawing up response plans without the agency's knowledge. A July 17, 2014, email by WHO outbreak expert Dr. Eric Bertherat described "total confusion" at the lab, which was then being shared by Metabiota and Tulane University researchers. Bertherat wrote that there was "no tracking of the samples" and "absolutely no control on what is being done." Sample tracking typically involves anonymizing samples by assigning them an identification number while creating electronic and paper records that provide identifying information about the individual from whom the sample was taken. Internal emails between WHO employees suggested that Metabiota did not have a system in place for sample tracking. Public arguments broke out between Metabiota staff and WHO representatives in Kenema. The AP investigation found that the intense feuding between different groups contributed to misdiagnosis of Ebola cases and repeated misreading of the trajectory of the virus.[2]

Metabiota chief executive officer and founder Nathan Wolfe said in a response to the AP report that his company specialized in tracking epidemics, not responding to them. He contended that some reports of mistakes were based on misunderstandings, while others were deliberately spread by competitors to undermine the company.[3]

Other pharmaceutical companies and governmental agencies had also set up in Kenema during the outbreak in an effort to learn more about the virus, but there was little to no collaboration among responders. There seemed to be greater interest in collecting data and concealing findings than in sharing information.

Working alongside the medical staff at Kenema and under the auspices of the MoHS, Gire and Yozwiak had already started sequencing Ebola virus genomes from 99 clinical samples collected from 78 Ebola patients in Sierra Leone within the first 24 days of the outbreak. The pair had shared the results of their initial sequencing results with the team in Kenema, as well as with Brima Kargbo, chief medical officer of Sierra Leone's MoHS. But after the WHO took control, their access to data was delayed.

Since the WHO was managing multiple individuals and organizations, some of whom the agency suspected of exploiting the chaos during the outbreak, perhaps it was difficult to distinguish the intentions of any particular group. In addition, there was no formal reporting structure for investigating and addressing suspected bad practices. As a result, the WHO and other agencies' response was to be prohibitive, even toward research units that had been in place before the outbreak. Restricting access helped agencies take stock of the situation and establish their protocols during a tumultuous time.

Gire and Yozwiak returned home on July 10 with the overwhelming feeling they were no longer welcome in Kenema. In the weeks and months that followed, they witnessed the devastation from afar. Nurses went on strike for unpaid wages. Anxious family members saw their loved ones enter Kenema Hospital only to die. Angry citizens believed the outbreak was just an excuse for foreign invasion and viewed the hospital as the site of a government-directed conspiracy. Rioters attempted to break through the hospital gates and rescue the patients.

Soon after Gire and Yozwiak left, Mbalu Fonnie, Kenema's head nurse and a prominent member of the community, was diagnosed with Ebola. On July 21, the day she died, while the clinical staff was trying desperately to save her life, nearly two hundred people gathered near the hospital with the intent of removing her from the Ebola ward. That same day, Humarr Khan tested positive for Ebola. A few days later, more Kenema residents, frustrated that Kenema was attracting Ebola patients from across the country and placing them at risk for infection, began rioting. Local police used tear gas and shot live bullets in the air to disperse the crowd.[4] Khan stepped outside the hospital on a few occasions to encourage the nurses and calm the crowds. Kenema became the setting for a power struggle between the local people, foreigners, and the virus, and Khan was caught in the middle.

So many people were devastated by Khan's death, but it motivated the team from the Sabeti Lab, now back in the United States, to pursue their efforts all the more strongly. They regarded their work on the Ebola virus genome as a way to honor Khan and direct more attention to the growing crisis. "Dr. Khan was a clinician through and through," Gire said, "but he understood how important the science was to patient care. He perfectly married science with clinical work."

Khan was so dedicated to his work that he had decided to stay in Sierra Leone even though that decision was a major factor in ending his marriage.[5] His wife, who was from Sierra Leone but was living in London, had begged him to move and practice there. The pair had met and fallen in love while Khan was in a medical residency program in the United Kingdom. They were married in Sierra Leone, but their wedding was the only time she agreed to return to the country. "I don't want to live in the bush," she would tell Khan. For more than a decade they lived out a long-distance marriage.

Khan did not want to abandon Kenema. He thought about his family, too. Khan's two children, born from two different women while he was in medical school in Sierra Leone, lived with his sister Mariama in Lungi. He would frequently make the nearly five-hour trip from Kenema on special occasions and on his way to the airport.

In 2014, the Islamic fasting month of Ramadan began at the end of June, and Khan had planned to spend the final two days, including the Eid al-Fitr holiday, in Lungi. But more and more Ebola cases were arriving in Kenema, and staff members were dying. "Mariama it's too hectic here," Khan told his sister during a phone call. "You will need to take the kids to pray with the parents." Days later came Khan's Ebola diagnosis. On July 29, after completing the special morning prayer on Eid, the children learned of their father's death.

Even as the staff at Kenema mourned, they continued to work relentless hours in the summer heat, donning and doffing their PPE. Foreign researchers and responders continued the power struggle. The virus became the backdrop of the outbreak. "It was a political nightmare," Yozwiak said. "And through this, we had this sequencing data emerging that was telling us something very important about the outbreak that was devastating Kenema."

The team had found more than three hundred genetic changes that make the genomes of the 2014 Ebola virus distinct from those of the viruses that had caused previous outbreaks. They also were able to determine from sequence patterns that this outbreak had started from a single introduction into humans, which subsequently spread from person to person over many months.[6] "We had estimates on when we thought the virus had entered Sierra Leone," Yozwiak said. "It was too early to tell anything about the mutations involved, but we were saying that if we keep analyzing the data we can monitor how this is mutating, and that's important for diagnostics, vaccines, and therapies."

The team had a choice. One option was to keep their findings to themselves until they learned more about the virus and its origin or were able to develop diagnostic tests or treatments over the next few years. A second option was to release their findings to the public, which would allow other researchers and drug developers to make use of them. They decided that in order to honor their colleagues at Kenema and to help the people still in harm's way, they would take the second option and make the data available to the global scientific community.[7]

Just days before the team was set to publish their results in *Science,* the journal received a cease-and-desist letter from Sierra Leone's minister of health, Dr. Miatta Kargbo, claiming that the work had not been authorized, despite the fact that multiple public documents both from the US government and Sierra Leone, including two provided by the healthy ministry, had authorized the sample sequencing and the procedure of sending samples to the United States.[8] Eric Lander, president and founding director of the Broad Institute, which housed the Sabeti Lab, contacted Dr. Sylvia Blyden, special executive assistant to Sierra Leone's president, Ernest Bai Koroma, to find out what was going on. She acted quickly, bringing in key public health officials, and rectified the situation.

The cease-and-desist letter was traced back to a low-level MoHS employee who was subsequently fired. President Koroma also fired the health minister, saying that she was removed "to create a conducive environment for efficient and effective handling of the Ebola outbreak."[9] In an email regarding the matter, a high-level official in the Sierra Leonean government noted, "I do not understand why any patriotic Sierra Leonean will want to stop this report from being published."[10]

On August 28, 2014, the *Science* article describing the Ebola genome appeared online (with the print version following two weeks later), authored jointly by teams from the Sabeti Lab, Tulane University, and Kenema Hospital. By that time, five of the paper's authors had died of Ebola, including Kenema's head nurse Mbalu Fonnie, nurse Alex Moigboi, and Sheik Humarr Khan, who had all been infected while providing care to patients.[11]

Although competition is an accepted feature of scientific advancement, the stakes are different during a global health crisis. In these situations, rivalry inhibits scientific and clinical progress, while information sharing and collaboration serve to advance science and accelerate the response, help stop the outbreak, and save lives. Yet the culture of outbreak response is so destructive that forces stronger than the pathogen impede the ability to contain epidemics. We found this to be true in past outbreaks, and it was confirmed by the results of our survey on the Ebola response. This survey gave us valuable firsthand information about the experiences of outbreak responders concerning collaboration and data-sharing.

Many respondents to the survey reported that they communicated well with colleagues within their own organizations, but they found communication and coordination with other organizations and with government agencies to be challenging. In response to a question about how decisions by international organizations were made, 21 percent indicated that they were made either completely or somewhat unilaterally. Some commented that there was no communication platform where individuals or organizations across agencies faced with similar situations could share their experiences and knowledge. For example, responders in all three countries experienced resistance from the local community, but no unified initiative was put together to coordinate an effective response. When asked what could be done to create a more collaborative environment during an outbreak, three quarters of the respondents recommended that organizations should work together—under uniform policies—to make decisions.

Many respondents to the survey who worked in Sierra Leone indicated that lack of communication and coordination between agencies—international NGOs, government agencies, and UN agencies—was a major barrier to tasks such as data collection and clinical care. They found this aspect of outbreak culture to be a greater barrier to their work than the other options listed in the survey question, which included hostility and resistance from local people, and government restrictions from the host country. Confusion about which agency was in charge of the response, and battles among agencies about which one would serve as a centralized data source, sparked public conflicts that "led to a delay in response activities, duplication of activities and lack of communication of important information," wrote one respondent. "The vertical pillars made sharing of information quite difficult, and there really was no centralized source who had a grasp on the overall response and where gaps existed or further inputs were needed," wrote another.

In Liberia, which had the second largest number of cases, the situation was not much different. The lack of coordination between the Liberian government and outside agencies led to what one respondent summarized as "too many cooks in the kitchen." Another wrote, "It was clear that agencies such as WHO, OFDA [Office of US Foreign Disaster Assistance], World Bank, Unicef, etc were simultaneously professing an intention to work 'hand-in-hand' with the government of Liberia while also making independent decisions about priorities for supporting the response." The same respondent continued, "The discrepancy between professed leadership and actual work each agency was undertaking created a chaotic and frequently duplicative and inefficient environment." Another person reported that the lack of coordination was, in part, a result of individuals and agencies working with the intent to enhance their own agendas: "Initially, it was difficult as everyone was concern[ed] about their turf and getting the glory."

One respondent remarked specifically on the poor relationship between larger and smaller agencies, describing the situation as "parallel processes of large NGOs undermining the more successful, scaled and cheaper interventions of smaller NGOs."

Despite respondents' complaints about the level of collaboration, many of them acknowledged that individuals, agencies, and organizations did work together because they had to. They reported that it became somewhat easier to do so as the months progressed. In Sierra Leone, the government, international NGOs, and the local community eventually came together to innovate, compromise, and devote all their efforts to the outbreak response. It just took too long and cost too many lives to get the response to the level it needed to be.

One of the most difficult tasks during any disease outbreak is gathering information. We asked several questions about this topic in our survey and learned that at the beginning of the outbreak, people working in all three countries found it extremely difficult to access relevant information, such as the numbers of new Ebola cases. Most respondents gathered information from a variety of sources, including on-the-ground experience, NGOs, government agencies, and their own colleagues and employer.

Information about the severity of the outbreak and about what was happening on the ground were hard to come by for many reasons. The most difficult types of information to access early on were data on clinical symptoms and on variation in the virus's genome, as well as ongoing scientific studies on other aspects of the virus—including clinical therapies for prevention and treatment—and information on patient cases and tracking of the virus's spread. One reason many international responders had difficulty assessing the situation was that they only stayed for a short amount of time. New protocols were put in place by each group, and by the time they implemented a working system or got to know the local medical staff, it was time to leave. One respondent wrote: "CDC should not provide six-week contracts during an outbreak response. It takes two to three weeks to understand complex country dynamics, add a field trip, a few hundred emails and the staff person leaves. Proper handovers to partners and to incoming staff did not seem to occur. Working with the CDC 'Ebola tourists' was extremely frustrating. Contracts should be three to six months for continuity." These short tours also decreased the opportunity for international responders to make connections with the local community. Local responders indicated that they did not know who some of the foreigners were or what they were doing. Some people within responding agencies collected information and took photos of individuals without their consent, one respondent wrote. This only extended the wedge of suspicion that local people had of foreign agencies.

Three levels of data are usually collected during an outbreak to help evaluate its severity and trajectory. Individual-level data provide detailed clinical information about specific cases. Exposure-level data involve information about specific events, such as funerals, or places, such as a particular household or workplace, which may have led to transmission. Population-level data describe the characteristics of a group of people or their environment and may help to understand why and track where the virus is spreading.[12] In an ideal environment, a centralized source such as a public database that contains information on all three levels of data—collectively contributed and routinely updated by responding agencies—would help responders develop a timely understanding of an outbreak. Such a centralized source for information would also help improve data accessibility for all responders and would improve coordination and collaboration during a response.

Survey respondents noted that the ability to find and share clinical data varied throughout the epidemic. During the period of March to October 2014, it was easy to find case numbers (the number of cases of Ebola) but harder to find clinical study results. This is probably because during this time, the health ministries of the West African governments were reporting case numbers, but little clinical research had yet been completed. It wasn't until September 2015, more than a year after the initial outbreak and as new cases were starting to wane, that clinical study results became more readily accessible.[13]

While difficulty in accessing data affected all types of outbreak responders, it was especially problematic for local clinicians caring for Ebola patients. By 2017, more than a year after the epidemic was declared over, clinicians still lacked standardized data that would help them diagnose the disease when a person first entered a clinic.[14] During the height of the outbreak in 2014, ETUs across the three countries lacked a standard protocol for diagnosis and care; informational material at different locations contained discrepancies in everything from the list of possible Ebola symptoms to how to manage patient care.

One of the recurring problems during outbreaks is the question of which group will "control the data"—that is, have authority over its input, maintenance, and accessibility. Our survey asked respondents to select the most significant barriers to data collection (with options that included competition, government or employer restrictions, and hostility from various sources) and also to indicate which parties helped or hindered their access to data. Several respondents reported that competition for intellectual and financial rewards made data collection impossible; others pointed to difficulties caused by problems with communication between agencies, and resistance from local people. Restrictions from employers, such as limits on traveling or the period of time allotted in a country, also posed obstacles to collecting data on the ground. Some respondents reported that they were hindered from collecting data by the governments of their own countries and local governments, as well as by industry staff members. Most often identified as aiding in data collection were colleagues, aid workers, and individuals from NGOs.

Data hoarding also proved to be a major barrier. Data hoarding includes the intentional suppression of information by an individual, group, or agency through tactics such as secretly collecting data, refusing to share data, and delaying the release of data, for personal gain. Laura Merson, associate director of the Infectious Disease Data Observatory at the University of Oxford, summed up the situation: "People were just collecting what they could. [Non-governmental organizations] were keeping their data private; academics take a year to get it out; and West Africa had set up surveillance but they were siloed from the international systems."[15] Several survey respondents noted frustration when a new cohort arrived and collected the same data as their predecessors, probably because information was not shared with the incoming group. While there was no requirement to share information, this repetition of work slowed down progress to control the epidemic. Many anecdotes from responders suggest that the spirit of collaboration diminished during the height of the outbreak.

To find out more about the misconduct that occurred during the outbreak, we asked respondents whether they had experienced data hoarding, intimidation, unethical tactics, or illegal tactics during the outbreak. If they answered yes, they were requested to indicate which category they fell into: victim of, witness to, heard about, or perpetrator of the tactic. We also asked if the tactic was helpful or harmful to their work (Tables 3.1 and 3.2).

TABLE 3.1
Experience of improper behavior among survey respondents

		Mode of experience, % (no. experienced/no. responses)			
Type of behavior	**% who experienced (no. experienced/ no. responses)**	**Victim of**	**Witness to**	**Heard about**	**Perpetrator of**
Data hoarding	43 (52/120)	25 (13/51)	43 (22/51)	29 (15/51)	2 (1/51)
Intimidation	37 (54/147)	46 (25/54)	31 (17/54)	20 (11/54)	2 (1/54)
Unethical tactics	38 (52/136)	14 (7/51)	55 (28/51)	29 (15/51)	2 (1/51)
Illegal tactics	27 (36/131)	11 (4/35)	51 (18/35)	34 (12/35)	3 (1/35)

Of the 120 participants who responded to a question asking whether they had experienced data hoarding, 43 percent said they had. Almost all of them reported that they were victims of, witnesses to, or heard about data hoarding—only one person described themself as a perpetrator. Half of the respondents saw the practice as harmful to their work. This may be because they could not access information that was critical for their ability to conduct research or effectively track and respond to cases. Nearly a quarter, however, described data hoarding as helpful, perhaps because it enabled them or their agency better access to data.

More than half of the survey respondents strongly agreed with the statement that "outbreak data should be shared openly." Nearly a quarter said that they somewhat agreed, while 16 percent said that they did not agree very much or at all with this statement. Sharing information can enhance collaboration and help advance

TABLE 3.2

Percentage of survey respondents who regarded improper behavior as helpful or harmful to their work

	Data hoarding (*n*=92)	**Intimidation (*n*=99)**	**Unethical tactics (*n*=101)**	**Illegal tactics (*n*=88)**
Helpful or very helpful	24	31	30	31
Neither helpful nor harmful	26	35	36	35
Harmful or very harmful	50	33	32	34

Note: n= number of respondents who answered the survey question.

knowledge during an outbreak, but it remains challenging and potentially disadvantageous, and there is little incentive to do so. Individuals and agencies may also have limited capacity to share information during an emergency situation. Some data may be restricted for reasons of privacy and consent. Even if an agency wished to share data, there is no centralized source for posting or accessing it, or any governing structure to ensure legal and ethical use of information provided.

One respondent wrote: "CDC data managers operating in Liberia reportedly refused to share data for over 4 weeks with other groups. They also insisted on using a locked system for data management, which was not ideal for information sharing and planning of appropriate outbreak response." Agencies including the CDC may defend the decision to use a locked data system for purposes of privacy and protection of sensitive information, rather than as a method of preventing access. Since no single agency "controls the data," the ethics and legality of hoarding data are debatable. What is certain is that there was a notable lack of interagency collaboration despite the high stakes of this situation.

In addition to data hoarding, many respondents reported that they had experienced intimidation and unethical tactics (such as maleficence and injustice). Of the 37 percent of respondents who indicated that they experienced intimidation, 46 percent reported being victims, 31 percent were witnesses, and 20 percent heard about intimidation; only one person reported being a perpetrator. About the same percentage of respondents, 38 percent, reported experiencing unethical tactics. The majority of those (55%) were witnesses; 14 percent described themselves as victims, while again, only one reported being a perpetrator. In the case of both intimidation and unethical tactics, about one-third of respondents believed that the tactic was helpful to their work, while one-third found it harmful.

There are a variety of possibilities as to why intimidation and/or unethical tactics could be considered helpful during an outbreak. Perhaps such tactics were deemed necessary in the moment for gaining access to information that would be critical for protecting hospital staff or providing patient care. Or perhaps they enabled responders to bypass bureaucratic processes that were stalling progress. Both tactics, however, were also described as contributing to the toxic culture that emerged at all levels of outbreak response. These tactics could also be for selfish purposes; respondents described scenarios of some research groups intimidating other groups in order to gain primary recognition of research findings.

More than a quarter of respondents (27%) experienced illegal tactics, including bribery and criminal threatening. Only 11 percent reported being victims, a lower percentage than for the other tactics surveyed. Half (51%) witnessed illegal tactics, a third (34%)

heard about them, and only one person reported being a perpetrator. Results about how helpful or harmful these tactics were to respondents' work were similar to those for the other tactics: about one-third (31%) found them helpful, and one-third harmful (34%).

A number of respondents to the survey described their experiences with unethical or illegal tactics, which helps us understand what kinds of activities went on, why people engaged in them, and what effect they had. The enormous pressure that researchers experienced, from both internal and external agents, may have played a considerable role in the reported behaviors.

The harmful competitive environment coupled with the chaos of an infectious disease outbreak, one respondent wrote, led to "manipulation of data for publication when [the] study was not conducted." Another reported witnessing data collectors attempt to obtain specimens from patients "in an unethical way and without cultural sensitivity." Some said they also witnessed research, including clinical trials of certain therapies, being conducted without host government approval.

Other reports highlighted the misuse of equipment. Some workers took advantage of humanitarian funds by "renting" cars that did not exist, or buying fuel for private cars. "There was a misuse of donated equipment and material," wrote one respondent. "I saw staff who [were] in theory on pay by the international community but who would then not show [up] to work. The reaction of some local health staff and authorities, particularly in the early phases of the outbreak, was really unethical as they were clearly more interested in getting a cut on the funds than in actually ensuring an adequate response to the outbreak."

"I was witness to international agencies knowing deployment of substandard PPE that did not meet specifications to protect healthcare workers from infection," wrote another respondent, not specifying the name of the agencies or the brand of equipment. "These national staff, then, did not have access to high-level Ebola care if they were infected."

Additional accounts mentioned the misuse of equipment to provide preferential treatment to certain patients: "I heard of government supervisory health workers removing from the holding unit grounds instruments and equipment (e.g., spray equipment for disinfecting bottoms of shoes and boots), affirming that they had 'purchased' this equipment someplace else, brought it in for some unknown reason to the unit, and then 'taking it back home,' presumably to offer some service to people outside the unit, unauthorized." Others who reported similar incidents believed that the health workers were taking equipment to protect themselves and offering preferential treatment to selected families.

Several respondents saw local leaders leave their posts to protect themselves. This behavior, like many that occurred, was probably motivated by fear. One respondent wrote: "I witnessed government health worker(s) in a supervisory capacity who had basically left their posts during the epidemic, leaving subordinates to assume the risk and responsibility for caring for people with EVD [Ebola virus disease]." When the supervisors returned to their posts, the respondent reported, they would ostracize the subordinates, take them off work schedules, "and instead schedule relatives and friends to work in units where incentive pay was being provided by NGOs."

This account of health workers leaving their posts is similar to reports noted elsewhere, of local volunteers within clinics across the region avoiding patient care for fear of contracting the virus. Adaroa Igonoh, a Nigerian doctor who was part of the medical team that treated the first Ebola patient in Lagos, Nigeria, recounted volunteer doctors who "showed up, but evaded the responsibility conferred on them to help the sick." "They would be around, but would not respond when called by patients," Igonoh wrote in a 2015 editorial in a scientific journal.[16]

One survey respondent recalled witnessing a worker with the UN's World Food Programme who was tasked to deliver food to a house of quarantined people instead leaving the food with some other people on the way there. Such incidents undermined quarantine efforts, since those who were supposed to be confined had to leave their homes in search of food.

Another respondent who worked with a small international organization admitted that personnel within the organization deceived one country's government officials in order to send aid to villages in need. "We couldn't bring goods up country because we weren't an official INGO [international non-governmental organization]," the respondent wrote. "So our workers made t-shirts that said Ebola Response and then went to officials who were in charge of giving road passes and gave them some money to get approval to take the aid [from] our organization to the villages that had asked [for it]." Knowledge of this incident was reportedly kept from superiors. "I was told [about the incident] later, because I could not have approved this," the respondent concluded.

The Ebola epidemic was among the first outbreaks during which rigorous field research to analyze transmission routes and to test experimental therapies took place while the outbreak was still in full swing. These types of studies are extremely difficult to conduct during a large-scale outbreak, not only because the pathogen itself is continuing to evolve, but also because of the complexity of the environment. Research to test the efficacy of vaccines and drugs, for example, requires sufficient statistical power and informed consent, must be well controlled, and takes time to be conducted thoroughly.[17] These requirements are less challenging when there are large numbers of cases, yet the goal in outbreak response is to prevent cases from increasing. The difficulty of designing a proper clinical trial during an outbreak restricts such research. So, too, do bureaucratic hurdles. Setting up clinical trials within host countries requires mounds of paperwork, meetings, and approvals that can take months, at best. In the case of pathogens for which there are no existing vaccines or treatments, like Ebola, those months are critical to saving lives. During a fast-moving outbreak, the pathogen inevitably outpaces the research.

Competition among researchers is fueled by time pressure and high stakes. Trudie Lang, professor of global health research at Oxford University, wrote in an editorial in *Nature* that the enormous level of bureaucracy and competition she and her team experienced while conducting a trial in Liberia on an antiviral drug slowed their progress. Her team planned to work with an experienced African staff, but after recruiting the staff in October 2014, learned they could not secure visas for them. The team also experienced delays in contract drafts and approvals to conduct the work. "Just as the epidemic began to show signs of slowing, we were delayed by six crucial weeks while waiting for contracts to be processed through MSF's systems, which took longer than seemed necessary," Lang wrote.[18]

By the time her team and four other research groups finally began research at the end of 2014, they had little time and few patients. Multiple teams vying for access to the same patient data in countries with inadequate health systems contributed to the chaos on the ground, according to Lang. "It was ludicrous," she said. "Because essentially we all had to fight over the same patients. It was like a land grab, and by that time the [new] cases were going down."[19]

Our survey confirmed that scientists found it challenging to conduct research during the outbreak, both in the field and in the lab. One respondent described "dealing with unprofessional colleagues" as a barrier, and another mentioned "colleague jealousy."

One researcher who responded to the survey reported that the worst experience during the outbreak was "losing months of work because of flaky contributors [and] having colleagues use our ideas without citation (and without letting us know they were even working on the problem)." Another, who reported having been involved in Ebola outbreaks for more than 20 years, wrote, "I have not seen so much greediness among researchers, most of these forgetting [the purpose of] the field [of research] besides using it for their own profit."

Almost everyone who participated in the Ebola outbreak response faced some kind of obstacle. Early on, the magnitude of the outbreak was downplayed, which led it to escalate. This was largely due to ignorance on the part of various agencies and local governments, which were reluctant to acknowledge existence of the disease. The county health departments that oversaw all the NGOs in that county often had the final say over what could and could not be done, which added to the confusion. There were also two major approaches to attacking Ebola: the conventional humanitarian response, and the politically and bureaucratically managed approach. The clash between these two approaches initially led to an uncoordinated process. Although these problems did eventually ameliorate, by that time the crisis was already underway.

THE WAVERING RESPONSE

My most profound regret is the WHO's slow response to the Ebola outbreak in West Africa.

—Margaret Chan, 2017, in an editorial reflecting on her tenure as WHO director-general

IN AUGUST 2014, one week after Sheik Humarr Khan died of Ebola, infectious disease physician Nahid Bhadelia boarded a plane in Boston on her way to Kenema, Sierra Leone. Bhadelia was joining a response team that had been formed under the WHO's Global Outbreak Alert and Response Network (GOARN). She had never seen a patient with Ebola.[1]

Bhadelia had been hired in 2011 to lead the medical response program for Boston University's National Emerging Infectious Diseases Laboratory, one of a handful of biosafety level 4 (BSL4) laboratories in the United States.[2] Her position as medical director of the Special Pathogens Unit is part of the support system that the CDC mandates for institutions hosting BSL4 labs. Units like hers are meant to provide medical care in the unlikely event that a researcher becomes ill after being exposed to one of the deadly pathogens being studied. The lab was still under construction when the Ebola outbreak hit West Africa, and Bhadelia—who was still learning the ins and outs of one of the largest biohazard labs in Boston—was watching the next major threat emerge abroad. The Google alert she had set up years before for the terms "Ebola" and "Lassa fever" began pinging messages into her inbox with dozens of news stories daily.

The previous month, Bhadelia and George Risi, a Montana-based infectious disease specialist, had made plans to travel to Sierra Leone to research an uptick in Lassa fever. But that was before the Ebola outbreak. Now Risi, who had also never treated an Ebola patient, asked Bhadelia to join the team he was assembling under the WHO's GOARN to volunteer at Kenema General Hospital, the country's largest ETU. Volunteer teams were already in Kenema, but because of the high risk of exposure, they only stayed about six weeks before a new cohort took over.[3]

"By the time we got down there in August 2014, we thought things were getting better," Bhadelia said. "What we didn't know was that the situation hadn't changed and we were just going down there as their relief." Bhadelia and her colleagues were not the only ones who assumed as much. Numbers of cases had been fluctuating, and a recent drop suggested that the situation was improving. This is one reason a large-scale response was not launched. In Guinea, the decline in new cases during April and May 2014 was so significant that the CDC's team leader—a veteran virus hunter—headed back to Atlanta and told his colleagues at the agency that the outbreak was close to over.[4] One respondent to our survey reported witnessing a WHO representative "swanning around Liberia from March [to] July saying, 'I have been in many Ebola outbreaks before. There is nothing to worry about.'"

The virus may well have gone into a latent period. Or early containment efforts could have been working. But the apparent slowdown in the outbreak could also have resulted from poor flow of information and underreporting of cases. In March, the WHO and the Guinean health ministry documented two people known to have been infected with Ebola crossing into Sierra Leone. That information had not reached Sierra Leonean health officials and teams investigating suspected cases in Sierra Leone.[5] It wasn't until May that Sierra Leone recorded its first confirmed cases. Those cases linked back directly to the two March cases that had never been reported.

The situation that unfolded between March and August epitomizes the worst of outbreak culture. The chaotic environment led to arguments between individual responders as well as between organizations, some of which played out publicly. Much of the frustration during this time, we found from our study, had to do with false certainty about how to handle the outbreak, and lengthy negotiations that hampered quick action.

The WHO had been instructed by Sierra Leone's Ministry of Health and Sanitation (MoHS) to report only laboratory-confirmed Ebola deaths, which led to severe underreporting because people who died before reaching a clinic were not counted.[6] Staff from Médicins sans Frontières (MSF) had also approached the MoHS in Sierra Leone in April to offer help, but were told that the situation was under control. In contrast to the WHO, a global health agency with perceived authority and clout, MSF brings in a team of health workers to manage the clinical response and often works with local clinicians. MSF's interactions with patients and local health workers often enable its staff to gain insight into the severity of an outbreak early in the cycle. In this case, MSF initially accepted the WHO's assessment that the situation in Sierra Leone was manageable because it was already overburdened by its Ebola response efforts in Guinea and Liberia. Meanwhile, the virus was spreading undetected.

In global disease outbreaks, two main sources of delay often impede a quick response. The first is a delay in the detection of cases by healthcare providers, laboratories, and public health authorities. The second is a delay in mobilization of response efforts. During the Ebola outbreak, although both played a role, the second—deferred mobilization—seems to have been a greater source of delay than the first—poor surveillance and detection.[7] It also proved to be the difference between life and death. Beginning in September, nearly 3,000 treatment beds were added in Ebola holding centers, clinics, and ETUs across 12 districts in Sierra Leone. As many as 57,000 Ebola cases and 40,000 deaths may have been prevented within the following eight months because of these improvements. But an additional 12,500 cases might have been prevented had the beds been introduced just one month earlier.[8]

Despite being strapped for resources and personnel, MSF publicly voiced its concerns when it became clear that conditions were worsening. In June 2014, the organization declared Ebola "out of control" and called on international agencies to take heed.[9] Although this statement may have amplified awareness of the catastrophe, it did not result in a significant international response. The WHO disputed the severity of the outbreak and the dire need for containment.

Almost a year later, in March 2015, MSF discussed this delay in a highly critical report. The organization accused the governments of Guinea and Sierra Leone, along with the US biotechnology firm Metabiota, which had been hired by the Sierra Leonean health ministry to monitor cases, of deliberately obstructing its early efforts to respond to the outbreak. The report, entitled "Pushed to the Limit and Beyond," contends that Guinea and Sierra Leone downplayed the severity of the outbreak and criticized MSF for "scaremongering"; it also claims that Guinea's president, Alpha Condé, accused MSF of overstating the threat as a ploy to raise more money.

It notes that MSF had alerted the WHO and the Sierra Leonean MoHS that people infected with Ebola had entered Sierra Leone as early as March 2014, and yet Metabiota and Tulane University, which were conducting surveillance of suspected cases out of Kenema Government Hospital, "seem to have missed the cases of Ebola." MSF's emergency coordinator is quoted in the report as saying that in June, when MSF finally opened its clinic in Kailahun, Sierra Leone—where Khan eventually went when he became infected—Metabiota and Tulane "refused to share data or lists of contacts with us, so we were working in the dark while cases just kept coming in." The report concludes, "For the Ebola outbreak to spiral this far out of control required many institutions to fail. And they did, with tragic and avoidable consequences."[10]

The release of the MSF report sparked a Twitter feud between MSF employees and a WHO spokesperson about who was to blame for the slow response to the epidemic.[11] This instance of assigning blame is a textbook example of the spillover of outbreak culture into the public sphere. The WHO maintained that it stuck to its legal framework for declaring a public health emergency. Metabiota, too, contended that the company had followed regulations and was not authorized to share any of its results in Sierra Leone with any party besides government health authorities.[12] Tulane never publicly responded to the accusations, but Tulane virologist Robert Garry privately reached out to MSF officials to clarify the institution's position and defend it from what he believed were unfounded claims. He received multiple email responses from a high-level public health specialist at MSF acknowledging the agency's positive working relationship with Tulane and apologizing if the report seemed "troublesome." According to one of the emails, the MSF member said he would look into "what might be initiated on our end and whether we might find a fence-mending procedure to initiate."[13]

Whatever the cause of the delay in response, it had led to the chaotic situation that Bhadelia found when she arrived in Sierra Leone on August 12, 2104. Four days before, the WHO had declared Ebola a "public health emergency of international concern."[14] This declaration drew sharp criticism from the international community for having come too late. The Associated Press obtained emails of internal documents suggesting that the WHO had been holding back on declaring a public health emergency for political reasons. The agency feared that such a declaration "could anger the African countries involved, hurt their economies, or interfere with the Muslim pilgrimage to Mecca," according to the AP report. WHO experts discussed declaring a public health emergency in June, but, according to the documents, a director mentioned doing so only as a "last resort." The WHO has denied that declaring an emergency earlier would have had much effect on the course of the epidemic.[15]

The WHO's declaration was not an invitation for public health agencies to respond, as most agencies already had a presence on the ground. It did serve as official recognition by the WHO that there was considerable risk for an Ebola pandemic and that containment of the current epidemic would "require a coordinated international response."[16] But there is no written outline for what constitutes a coordinated international response.

Nearly two hundred countries have signed the WHO's International Health Regulations, an international treaty for managing infectious disease outbreaks that requires countries to develop core capacities to prevent, detect, and respond to outbreaks, and to quickly report outbreaks to the WHO.[17] Government health agencies, including those of the three affected countries, are ideally at the forefront of building such capacities and should use scientific principles of infection control to deploy public health measures, including trade or travel restrictions. Seven major reports that reviewed what went wrong during the Ebola outbreak and how infectious disease outbreaks should be better managed all identified inadequate compliance with the WHO's health regulations as a major contributor to the slow response to Ebola.[18] Government health agencies also lead efforts of public health messaging that spread awareness of an epidemic and educate citizens about how to prevent infection or treat the disease. In addition, they are responsible for approving any international responders and working with larger agencies to devise a response plan.

With the host government's approval, the WHO is primarily responsible for coordinating the overall response during a public health emergency, since it has the authority to manage health crises by diverting financial, human, and logistical resources to epidemic response. Though headquartered in Geneva, Switzerland, the WHO has autonomous offices in each of the six regions of the world. The agency also regularly publishes data about the status of an epidemic along with "situation reports" that summarize the epidemic's progression and discuss any containment challenges. However, its budget had been cut after the 2008 global recession. At the time of the Ebola outbreak, the agency had already been responding to numerous other emergencies and outbreaks, including MERS in Saudi Arabia, polio in Syria, and avian flu in China. Just six months before the young boy died from Ebola in Guinea, the WHO infectious disease arm underwent a $72 million budget cut, and its budget for "outbreak and crisis response" was cut by 51 percent.[19] In addition, the governing structure was fraught with bureaucratic overlap, politics, and confusion.

After declaring Ebola an international emergency in August, the WHO sent 50 people to the region to help with contact tracing, disease surveillance, laboratory work, logistics, information-sharing, and social mobilization, but did not send any clinicians or equipment.[20] The Africa office, based in the Republic of the Congo—

operating on a budget half the size it had been five years earlier, when it laid off 9 of its 12 emergency response specialists—led the early Ebola response on the ground.[21]

Another international agency that is involved in disease outbreaks is the UN, which has a much broader role than the WHO. The UN provides technical and logistical support by creating coalitions that include a member from each major responding agency along with representatives from the affected area. These coalitions hold weekly meetings to discuss whether operational needs are being met. Subgroups within the UN, including the World Food Programme and Unicef, also support the response by distributing food and health supplies and building up infrastructure.

Several other nonprofit and nongovernmental organizations also play a part. Some, such as Save the Children, International Medical Corps, and Partners in Health, build and manage clinical treatment units while also training local healthcare workers. Others, such as Oxfam and International Rescue Committee, provide infrastructure support and water and food aid.

For-profit research companies and academic medical centers, too, have a stake in outbreak response. They are often the ones collecting and analyzing samples of the pathogen and conducting clinical trials to understand its origin and behavior, as well as to develop therapeutics, if not already available. They work closely with government health agencies to gather data, devise trial protocols, and gain approval for clinical trials.

While each responding agency has a role based on its capabilities, these roles converge in times of crisis. The convergence does not always result in a seamless overlap. Instead, a clash of functions adds to an already chaotic environment. Agencies involved in containing the outbreak in West Africa were unsure of what a coordinated international response would look like. There was no outbreak response playbook, no unified approach, or any organized response from similar past outbreaks to follow. Each group of responders—governmental, nongovernmental, for-profit, or academic—acted as a separate entity with limited knowledge of what the other parties were doing.

Ebola was better controlled in the countries it spread to beyond the three core countries largely because of better coordination within their governments and less reliance on international responders. Nigeria, Senegal, and Mali, with their comparatively better health systems and swifter policy response, successfully contained flare-ups. Each country had its own high-quality laboratory to detect cases, and each used local workers and existing infrastructure to track and control cases.[22] While Guinea, Liberia, and Sierra Leone also utilized local staff, they relied more heavily on international responders and nonprofit organizations. However, some organizations, like the nonprofit International Medical Corps, delayed joining response efforts because they saw MSF's presence as a sign that the outbreak was being managed. "We didn't try to get involved earlier because there was a point a couple of months ago when it seemed like MSF is dealing with it and they have it under control," International Medical Corps' former deputy director Benjamin Phillips told Reuters in August 2014.[23]

Meanwhile, overcrowded MSF-run clinics had to turn away potentially infected people, denying them medical care and leaving them to die on the streets or in their homes. MSF had tried to sound the alarm on the emerging epidemic in June, saying the agency has "reached our limits."[24] The organization, which has an annual budget of over $1.5 billion, spent about $80 million on outbreak response in 2014.[25] It reported being overstretched in its response to Ebola, and it was responding to crises in Syria and Sudan at the same time. Calls for assistance had less to do with the potency of the virus than with the overall lack of resources. This outbreak, unlike previous ones, would require a massive deployment of resources by outside governments and aid agencies, according to MSF. "This was different and they [MSF] knew it was different," said Sheila Davis, chief nursing officer and the head of Ebola response at the nongovernmental organization Partners in Health.[26]

Even Partners in Health, which was acclaimed for its response to the 2010 cholera outbreak in Haiti, had a difficult time finding American healthcare workers willing to go to West Africa. Many felt disconnected from an epidemic that was occurring in a different hemisphere, Davis said. Most airlines stopped flying to West Africa, which created another disincentive for people to help.

It was Khan's death that sealed the decision for Partners in Health to go to Sierra Leone, according to the organization's founder, Dr. Paul Farmer.[27] Partners in Health workers arrived in the country on September 10, 2014, knowing they would be in it for the long haul. The organization maintains a presence in the country to this day.

On September 16, the WHO and Sierra Leone's Ministry of Health and Sanitation organized a three-day training program for nurses in the capital, Freetown.[28] The nurses learned about Ebola and about measures to prevent transmission of the virus. At the time, there were only two ETUs in Sierra Leone: one in Kenema and another in Kailahun. It was clear, however, that two ETUs were not sufficient, and that healthcare workers were in dire need of more intensive education.

The same day, MSF president Joanne Liu updated UN member states in Geneva on the status of the outbreak. It was the second time in two weeks that she had appealed to countries to send "more hands-on capacity" in the form of massive military deployments and medical personnel. The response, she said, was falling "dangerously behind." The United States announced it would send three thousand military engineers and medical personnel to Liberia to build clinics and train healthcare workers. But that was not enough, said Liu. The burden of containing Ebola was falling on nongovernmental organizations that did not have adequate funding, resources, and staff to launch a large-enough containment effort.[29]

Many of our survey respondents mentioned how frustrated they were with the lack of resources. "[The] biggest obstacle was delay in readily available funds to respond appropriately during the start of the outbreak which could have averted the scope of transmission," one respondent wrote. Another stated, "I was working at the district level, and we were so outmatched by the outbreak that it still takes my breath away to think about it. I was working in epidemiology, and there was an absence of people who were available and had the skills to go to the field to lead investigations and train local staff on how to do this."

The response was also hampered by health systems that were inadequate to begin with. Parts of the region's physical infrastructure and its health systems were already in shambles before the outbreak hit. Poor roads made it difficult for responders to reach many communities. Most hospitals and clinics were not equipped to handle minor illnesses, let alone cases of Ebola. A survey respondent who was working in Kono District in the Eastern Province of Sierra Leone wrote, "We watched in horror as the district government hospital was overwhelmed by the Ebola crisis in the fall of 2014 to the point where the Ebola patients in the hospital were uncared for, hospital staff were woefully undertrained and under-resourced to meet the need and bodies piled up in the hospital."

There were not enough ambulances to transport the sick to ETUs. Respondents said that in some cases, family members were not told where their sick relative had been taken. At hospitals, the lack of a sanitary waste management system exposed people to the virus who otherwise would never have gotten the disease.

Many people concluded that they had a lower risk of contracting Ebola if they just stayed home.

The insufficient response did more than cause the disease to spread. "This was not just an Ebola emergency," one respondent wrote. "It impacted greatly on normal health care, education services for children and subsequent academic learning, and also on child protection issues." "The barriers seemed systemic," wrote another. "Poverty is at the root of the problem, and those types of barriers can seem almost insurmountable."

"We need you on the ground," Liu pleaded to the UN member states in mid-September. "The window of opportunity to contain this outbreak is closing. We need more countries to stand up, we need greater deployment and we need it now." Liu read those words verbatim from a different address she had delivered to the UN member states in New York just two weeks before. Meanwhile, nearly half of the more than 4,000 people known to have been infected with Ebola in West Africa by that time had died.

Two days later, the WHO's director-general, Margaret Chan, briefed the UN Security Council, calling Ebola "the greatest peacetime challenge that the United Nations and its agencies have ever faced." The challenge was immense: "None of us experienced in containing outbreaks has ever seen, in our lifetimes, an emergency on this scale, with this degree of suffering, and with this magnitude of cascading consequences," she said.[30]

As a result of this briefing, the UN Security Council called on its member states to respond to the crisis and to refrain from isolating affected countries. This was the first time the council had ever officially commented on a public health event.[31] The directive against isolation heeded recommendations by the WHO's emergency committee on Ebola, which had met the month before. The committee recommended screening travelers from the three countries hardest hit by Ebola, but it believed that trade or travel bans would do more to harm than to help the response.[32]

Meanwhile, major hospitals in the three West African countries had little to no capacity to care for Ebola patients. After Khan died, the clinical team at Kenema Hospital consisted of four volunteer physicians from the United States, including Risi and Bhadelia, scores of volunteers who cycled in and out of the country, and only a handful of local nurses. A majority of the nurses were on strike because they were not receiving the hazard pay promised to them by the health ministry, or adequate PPE to keep them safe. In August, Dr. Sahr Rogers, a Sierra Leonean physician who had been working in Kenema, had also died from Ebola.[33] There were no local doctors left at the hospital.

The outbreak developed quickly while broader discussions of what to do, and how to do it, failed to keep up. In her final speech as WHO director-general in 2017, a year after the end of the epidemic, Chan acknowledged how poorly prepared the agency had been to handle Ebola. The WHO, she said, "was too slow to recognize that the virus, during its first appearance in West Africa, would behave very differently than during past outbreaks." Soon after, in an editorial reflecting on her tenure, she wrote: "My most profound regret is the WHO's slow response to the Ebola outbreak in West Africa."[34]

Chan's swan song for Ebola sounded much like then outgoing UN secretary-general Ban Ki-moon's speech on the handling of the 2010 cholera epidemic in Haiti, which had occurred under his leadership. In that case, the epidemic began because a UN team, which had gone to Haiti after the devastating 2010 earthquake, released untreated human sewage into the largest river system, contaminating the country's water source. For years, the agency denied setting off what is now considered among the world's largest cholera epidemics.[35] Six years later, after more than nine thousand people in Haiti had died of cholera, Ban Ki-moon acknowledged responsibility during a speech at the UN headquarters in New York City and apologized for the agency's role. "We simply did not do enough with regards to the cholera outbreak and its spread in Haiti," he said. "We are profoundly sorry for our role."[36]

Blame and regret, usually delivered retrospectively, are recurring themes in outbreak culture. The WHO often bears the brunt of international criticism during outbreaks, as it did during the Ebola epidemic, and to its credit it often does accept blame, express regret, and look inward for change. But it is not the only large-scale player involved in responding to outbreaks, or the only one that could benefit from introspection.

Other major agencies and governments, including the US government, also delayed their intervention in West Africa and thus contributed to the spread of the disease. Conversations among US government agencies revolved around how many people and also how much time and resources would be devoted to the crisis abroad. According to our survey respondents, US organizations were under the impression that they would respond in some way, but they were unclear about when and in what capacity. One of the respondents defined the wavering response by the United States as the "inability for policymakers to agree on the approach and framework to combat the virus." The largest obstacle, according to another, was "infighting within my US government agency about who 'got' to participate in the Ebola outbreak." This respondent continued, "There were more interested, able people who were not invited to the table."

Internal battles within international groups and among government officials (not just the US government agency just mentioned) were among the primary reasons there was limited outside response during July and August, a critical period of the outbreak. The majority of our respondents believed that if major international actors had undertaken wholesale action earlier—through sending personnel, medical supplies, money, and so on—it would have severely limited the opportunity for the virus to spread.

Academic medical centers in the United States reportedly discouraged staff from responding to this outbreak, unlike in previous epidemics, when staff were not only allowed to deploy to outbreak hot zones but were given incentives to volunteer, including time off and additional pay. The CDC issued a Level 3 travel warning, the highest alert possible, urging all US residents to avoid nonessential travel to Guinea, Liberia, and Sierra Leone. Academic centers also issued their own travel bans for staff, citing the lengthy training period required, as well as a lack of capacity to handle one of their own who might get infected with Ebola.[37] Some specified that if their staff wanted to help, they had to find a way on their own. Infectious disease experts and clinicians, even individuals experienced in patient care and outbreak control, had difficulty finding other organizations that would deploy them. Some of them submitted volunteer applications with the largest responding aid groups and received a response only months later, or no response at all.[38]

While healthcare workers in the United States looked for groups to connect with, a large cohort of personnel arrived in West Africa from an unlikely place: Cuba. That small country sent nearly five hundred healthcare professionals beginning in September 2014, the largest number of trained medical workers from any one country sent in response to the Ebola crisis. Cuba's focus on population health within its own country, as well as the training its workers received in responding to earthquakes and other major international catastrophes, uniquely positioned the country to battle outbreaks. Only MSF sent more clinical responders to West Africa at that time.[39] Soon after, the United States, the United Kingdom, Japan, and India followed with military contingents, money, equipment, and comparatively smaller teams of health workers.

The wavering response from major countries, both intended and unintended, exacted a traumatic toll on West Africans, especially those directly involved in outbreak response. Many did not care about the reasons why supplies and personnel weren't arriving, only that they were not arriving fast enough. When asked to describe their worst experience during the outbreak—besides witnessing Ebola cases—one respondent to our survey wrote, "realizing the world was waiting for West Africans to die off."

By October, some initiatives organized by international responders were helping to limit transmission of the virus. The United Nations Mission for Emergency Ebola Response (UNMEER), the first-ever UN emergency health mission, which was established in mid-September, was among the most effective interventions in Sierra Leone. Part of UNMEER's strategy was to actively find and isolate cases, trace contacts, and perform safe and dignified burials for people who died of Ebola. UNMEER was most successful in interrupting transmissions within households, one of the highest-risk places to contract Ebola. One model estimates that the average number of infections by transmission decreased by 43 percent after October 2014, and by 65 percent by the end of December, mostly due to case isolation and safe burials.[40]

Still, organizational glitches, including miscommunication and lack of coordination among international governments and agencies, continued. In October, the Russian military sent a group of specialists to establish a hospital in Kindia, Guinea—about 85 miles north of the country's capital, Conakry—and trained local healthcare workers to staff it. Despite agreements between the two governments, when supplies and personnel arrived, "there was nobody ready to unload the planes, refuel them or carry out other necessary administrative procedures."[41] The hospital opened in

December 2014, in large part because Guinean and Russian troops provided protection against theft of supplies and community resistance.

Meanwhile, within already established clinics in the three hardest-hit countries, healthcare workers were doing the best they could with woefully inadequate resources. They and their patients suffered the most as a result of the world's inconsistent response. Local nurses and doctors became interchangeable throughout the region, because there were so few. If a worker could start an IV, push fluids, and clean patients, their title was irrelevant.[42] It was no different in Kenema.

The staff at Kenema Hospital had put in place a triage system intended to separate suspected from confirmed cases of Ebola. But with an overcrowded and underequipped hospital, and rampant denial and mistrust, this was impossible to effectively do. The ward with suspected cases became just another breeding ground for the virus. Early symptoms of Ebola are similar to those of other diseases, so someone who showed up at Kenema with a fever and chills could have malaria, or cholera, or yellow fever. When the sick first arrived, they entered the so-called suspect ward—a concrete waiting room with cots and chairs filled with people who might have Ebola. They were held in the ward for up to three days, because that was how long it usually took for the results of blood tests to come in. If a person tested positive for Ebola, he or she was moved to the confirmed ward with other Ebola-stricken patients. If the test came back negative, the person was sprayed down with bleach water and sent home. But even if a person in the suspect ward didn't have Ebola, someone else in the ward might. For several days, anyone who came in seeking medical care was at a higher risk of contracting the fatal virus than if they hadn't sought help in the first place.[43]

Staff levels, supplies, diagnosis, and treatment were only a few of the resources that were lacking in Kenema. Harsh working conditions limited bedside care of Ebola patients. Staff members were ideally limited to one-to-two-hour intervals of working in full PPE under extreme heat and humidity, though the massive influx of cases made it difficult for them to abide by this protocol.

Field laboratory testing for Ebola relied on a cumbersome, complex, multistep process that had a long turnaround time and was not very accurate.[44] It typically took up to 72 hours to get results from the molecular diagnostic test that was being used for diagnosis, since blood samples from most ETUs had to be sent to an off-site laboratory. The test only worked when blood samples were taken from patients who already exhibited symptoms of the disease, which made it difficult to decide which individuals to isolate early on.

Despite the fact that six laboratories had been set up in the entire country to run diagnostic tests during the critical months of the epidemic, Bhadelia found when she returned to Port Loko in November 2014 that it took the same time to get test results back. Two Sierra Leonean phlebotomists hired by the district's health ministry were reportedly the only ones drawing the blood for testing. The pair traveled on a motorbike they shared from one suspect ward to another, drawing hundreds of blood samples. If their bike broke down, or they couldn't afford to pay for gas—which was considered an out-of-pocket expense—they just stayed home, potentially delaying sample delivery and testing for days.[45]

Even after two months of a fast-moving outbreak, not much had changed, even in the most heavily hit areas. "Billions of dollars later, many agencies on the floor, and we are still propagating the epidemic," said Bhadelia.[46]

DISTRUST IN A CULTURE OF COMPASSION

Living in fear and terror.
—Words of Liberian nurses and midwives caring for Ebola patients

AUGUST 2014 WAS one of the darkest months in the Ebola ward of Kenema Government Hospital. That month, workers in Kenema noticed that something was wrong with the chlorine powder mix they had been using to wash down the ward. Jerome Souquet, a WHO consultant, tested the batch and found there wasn't enough active ingredient for proper surface disinfection. "I was deeply shocked," he wrote in an email to a WHO official in Freetown. The effect, he said, "could be catastrophic, and cause immediate infection of all the staff."[1]

Barrels of the powder, which came from the government's supply in Freetown, had been expired for more than a year. Although hospital staff stopped using the defective chlorine, for days they were left with no alternative. Later it was found that there were also defects in the personal protective equipment (PPE), including surgical gowns and full face masks, that had been brought in from the United States and worn by many local and foreign health workers across the region.

One of the worst parts of the outbreak, in the words of one of our survey respondents, was "realizing that hundreds of thousands of Ebola PPE were deployed across an entire country but [with] substandard specifications. [PPE providers] tried to justify their use not based on evidence that proved their effectiveness, but a lack of data to say that they were not safe." The same person concluded, "There is no way to estimate how many Africans contracted Ebola as a result of this decision."

The risk of contracting Ebola grew with every second, minute, and hour a person stayed in Kenema's Ebola ward. Fluid-drenched patients lay in an overcrowded and understaffed facility, and the dead cluttered the hallways. Just weeks after thousands of local people gathered outside threatening to burn the hospital down, health workers went on strike citing dangerous conditions, strained resources, and delayed pay.[2]

Kenema's nurses, including head nurse Issa French, had been working shifts of 12-plus hours while grieving Dr. Sheik Humarr Khan's death. French had been part of the clinical team that accepted Kenema's first Ebola patient back in May. Within three months, French had cared for more than five hundred people who had been infected with the virus. The WHO has estimated that more people became infected with Ebola in August than in all the earlier months combined.[3]

Boston University physician Nahid Bhadelia worked alongside French as they struggled to provide supportive care to the dying. When the day was over, Bhadelia and the other foreign healthcare workers headed back to their hotels to rest. Many of the local workers, who had been forced out of their homes and rejected by their villages, slept in offices at the hospital. Some, like French, would remove their uniforms and hide any item that included medical insignia indicating they were Ebola workers for fear they would be shunned by their community.

Every day, for months, French worked tirelessly—and often futilely—to save lives. He understood the reality of the disease and the urgency of containing the outbreak better than most West Africans. The burden he bore did not go unnoticed. "Why do you risk your life?" Bhadelia asked him one day when he stole some time outside, away from the overcrowded ward. "Because this is my community," he responded. "Because these are my people. Because these are human beings."[4]

No matter how hard they worked, outbreak responders like French were no more than bandages to a people that had been deeply wounded by colonialism, corruption, and conspiracy. The recent history of colonialism and civil war in the region had fed skepticism about the story that a virus was the cause of the outbreak. The fact that key governments with colonial links to the three affected countries became central to the Ebola response added to the disquiet. France established treatment centers in Guinea and helped train healthcare workers. The United Kingdom provided funding and emergency, medical, and military support in Sierra Leone. The United States sent medical supplies and staff to Liberia. The legacy of colonial rule and postcolonial conflict hindered communication between local people and those responding to the outbreak. Guinea, Liberia, and Sierra Leone had experienced years of oppression and inequality, along with both foreign and domestic exploitation of their region's resources. Given this history, the distrust during the Ebola response is not surprising. Many citizens were suspicious of government motives and developed their own interpretations of the outbreak's origin. International health groups that had pulled out of West Africa during the civil wars hesitated to reenter. When these groups returned, their intentions were met with doubt.

The fast-moving outbreak did not give the limited number of foreign responders enough time to fully grasp the anthropology of the region. As a result, they communicated in ways that were often not effective. Prohibiting cultural practices such as some aspects of burial traditions only fueled the conspiracy theory that something man-made was behind the epidemic. Many people did not believe that Ebola existed. Some thought their government had conjured up fear of a plague to keep citizens in line. After all, a revolution taking place during an infectious disease outbreak would be less suspicious. Some thought it was a government-perpetrated genocide. Perhaps healthcare workers, whose faces remained hidden behind masks, were in on it too. Standard infection control practices certainly seemed suspect. When a person was confirmed to have Ebola and taken to the hospital, a team of health workers was dispatched to the home to burn their possessions, sometimes even the home itself. Patients discharged from care returned to nothing. Thousands entered the hospital, and most of them never came out alive.

In the initial stage of the outbreak, medical professionals and government officials tried to keep the public updated, but they often passed along unverified information, some of which turned out to be false. This, coupled with the spread of inaccurate news reports, made it very difficult to contain the disease. Across the region, even those who believed that Ebola existed had a hard time distinguishing facts about the disease from rumors. In Guinea's capital, Conakry, a rumor that eating onions and drinking a coffee mixture could protect against Ebola led to onions selling out in the city's main market. In parts of Sierra Leone, word spread that the disease could be cured by consuming a strong alcoholic drink called bitter kola. In Nigeria, where the outbreak was best contained, rumors that drinking salt water would protect against the virus led some to die from salt overconsumption.[5]

Liberia had fewer Ebola cases than Sierra Leone and Guinea early in the epidemic, but it was no less vulnerable to rumors and resistance. The first two Ebola cases in Liberia were confirmed in the northern town of Foya in March 2014, eight days after the outbreak was declared in Guinea. Some accused Liberian president Ellen Johnson Sirleaf of intentionally poisoning citizens through the drinking water and exaggerating the scale of the epidemic in order to receive international donor money. People thought the hospitals were a more dangerous environment than the villages. "They believed if they went to the ETU they would inevitably die," said Darlington Jallah, head nurse at the MSF-operated Foya ETU in northern Liberia. "They just ended up dying in their houses."[6]

Jallah cared for the first Ebola patient in Liberia, a woman who had arrived at the hospital in March from Guinea. Jallah had read about Ebola in his study of tropical diseases, but neither he nor any of the hospital staff had ever cared for an Ebola patient before. This patient was the hospital's first Ebola death. After that, waves of people overwhelmed the staff, even as foreign help arrived. At one point, according to Jallah, there were 35 patients and only 10 beds.[7] Those beds never opened up. Despite the chaos, Jallah managed to learn the name of every patient treated at Foya's ETU. To this day, he can recall each by name, and he has memorized the names and headstone locations of the 251 people buried at the nearby Ebola Memorial Cemetery. "I see the names on the gravesites, and I can see each of their faces—even the way they were lying in their beds," Jallah said.[8]

Foya district, part of Lofa County, is considered Liberia's food basket, as it is a major center for agriculture and supplier of food across the country.[9] Near the border of both Sierra Leone and Guinea, Lofa County was one of the areas of Liberia that had the highest number of Ebola cases.[10] Foya, a town of about twenty thousand, had been caught in the crosshairs of civil unrest during the 1990s, when fighters from both neighboring countries hopped the border to fight, essentially cutting off the food delivery system and crippling the county's economic base. Humanitarian organizations like Samaritan's Purse had been working for years on sanitation, agriculture, literacy, and child protection campaigns in Foya. The Ebola outbreak put a halt to the community's years-long rebuilding efforts. Once the first Ebola patients arrived in the town, Samaritan's Purse helped to manage the hospital's ETU, while simultaneously creating awareness and education campaigns to combat the doubts and conspiracies. Jallah and other local health workers battled community suspicions of the outbreak as they cared for patients dying of Ebola. "At the start of the outbreak, there was 90 percent denial that Ebola was in Foya," Jallah said.

Early public health campaigns by Samaritan's Purse and others initially centered on disproving doubts by posting signs that declared "Ebola is real." Unfortunately, this simple counterstatement did not work. Liberians also seemed unreceptive to signs telling people to go to the hospital if they felt sick, not to eat bushmeat, and not to attend funerals. Social mobilization to demonstrate Ebola's existence proved more effective than signs, probably because it played into the region's longstanding oral tradition. The strategy included deploying trusted community members to communicate behavior-changing messages.[11] Radio stations aired public service announcements in the form of catchy tunes and talk show segments. Pastors preached entire sermons about the disease to their congregations. Community volunteers backed by governmental and nongovernmental organizations went door to door as part of outreach and education campaigns. The transparency and community integration of these campaigns eventually paid off, and behavior began to shift.

It took months to establish trust and instill safe Ebola containment practices among the local people in Foya district, but by late August 2014 Liberia's Ministry of Health and Social Welfare reported a decrease in new cases of Ebola in Lofa County, suggesting that its citizens had begun to understand the severity of the outbreak and adhere to the public health messages.[12] That is, until one case collapsed the containment structure they had worked so hard to build. Her name was Pauline.

On September 19, an eight-year-old girl named Pauline arrived at Foya's ETU with fever and dehydration. Pauline had lived with her family in a village in Lofa County, but she was orphaned after her entire household died from Ebola. The family home had been burned down, a protocol meant to inhibit the spread of Ebola. Pauline moved in with her aunt Princess, a nurse at Foya's ETU. Princess recognized the symptoms immediately and brought Pauline to the ETU. Pauline's test for Ebola came back positive. But four days later, her symptoms had vanished. Two additional blood tests were ordered, and both came back negative for Ebola.

Had she been cured? Recovery was rare at that point in the outbreak, but it was not unheard of. Jallah grappled with the question of whether to discharge Pauline from the ward. He had never before seen such a quick recovery. However, with a negative test result and no symptoms to treat, Pauline was taking up space in the ward that could be better used by someone who needed it. The doctor printed a paper certificate identifying Pauline as a survivor, signed it, and told her she could go back to her village.

On September 23, despite pushback from local staff advocating for further monitoring, Pauline was sent home.[13] A team was sent to the little girl's village to help with her transition. As part of the hospital's outreach effort, health educators had spent months telling the community that anyone who was given a certificate was Ebola-free. The certificate became a survivor's right of reentry back into the village. The educators had explained that a survivor could no longer be infected by Ebola and could not pass the virus on to others. Having a certificate helped ease the stigma for survivors as they transitioned back into their community.

Debbie Wilson, an American nurse volunteer with MSF, arrived in Foya a few days before Pauline was discharged. Wilson heard about Pauline from Jallah and the other nurses. Pauline was Foya's success story, living hope in a ward that seemed to see only death.[14]

Wilson, who lived in Massachusetts, was part of MSF's clinical response team that had been sent to Haiti for three months during the post-earthquake cholera epidemic in 2010. The level of devastation and the sense of urgency there, she says, was nothing like what she experienced in Foya. MSF sent her to Liberia with a small group of international healthcare workers to set up a multi-tent, 120-bed Ebola clinic in Foya, and to train and equip 78 local nurses and a dozen nursing assistants. Unlike in Haiti, Wilson could not spend a three-month period in Liberia. Because of Ebola's virulence, MSF adopted standards set by the WHO to allow foreign healthcare workers to spend only six weeks in the country at one time. Workers had to wait at least 21 days before returning. The revolving door of healthcare workers frustrated the local staff. Every few weeks a different foreign team came in with a new way of doing things, often implementing procedures with no input from the local staff.

Wilson, along with another expatriate nurse, oversaw the local nurses during her time in Foya. She brought them with her to management meetings. She asked them about past practices and ensured that they were informed and involved in every stage of determining policies and practices in the ETU. One of the nurses once told Wilson of the nickname the staff had given her. "Do you know that we call you the 'Mother of the Blacks'?" the nurse said. The nurses told Wilson that for the first time since the outbreak, they

Debbie Wilson, American nurse with MSF (*kneeling, right*), with local nurses at Foya Hospital, Liberia. Head nurse Darlington Jallah is standing behind her in a white shirt. *Courtesy of Deborah Wilson.*

felt they had regained control of their hospital. Amid the chaos, Wilson listened when they spoke.

When nurse Princess arrived for her shift on September 29, she brought with her a frail and fatigued little girl. Wilson recognized the little girl as Pauline, who had left the ward just a few days before as an Ebola survivor. Wilson's eyes met Princess's at the desk in front of a triage room of suspected Ebola cases.

"Look, it's probably something like malaria," Wilson assured Princess. "Bring her in, let's check her out."

Blood and saliva tests both came back negative for Ebola. Another diagnostic test, this time looking at antibodies produced by the body's immune system in response to the virus, was sent for analysis to a pop-up laboratory nearby. Days later, during an administration meeting, Wilson learned the results. "Pauline has Ebola, but we don't want the others to know," an MSF administrator told Wilson.

"The others" meant the healthcare workers, Princess, and local Liberians. Outreach teams had worked so hard to eliminate the stigma of Ebola survivors in Pauline's village. They had told villagers that because Pauline was a survivor, she would never get Ebola or infect anyone else. The lab, which still had Pauline's blood work from her first admission, retested the previous samples and all of them came back negative, meaning there had not been a mixup of the results before her discharge. Pauline's relapse baffled the hospital administration, who feared that revealing it would confuse the staff and destroy the trust that had been built in the community.

This call for secrecy infuriated Wilson, and she refused to accept it. She only had a few days left in Liberia, so she knew she would not be there to see the outcome, but she didn't care. "Our nurses need to know; they need to protect themselves," Wilson told the administrators. And they needed to know immediately.

Wilson thought that Jallah, the head nurse, should be the one to tell Princess and the other staff members, since he would be there long after the foreign workers left. Wilson found Jallah standing next to a fence outside the hospital. As she walked over to him, he could sense that she was not just coming to tell him that another patient had died. She told him about Pauline's apparent relapse. He lowered his head in despair. As much as it pained Wilson to tell Jallah, Jallah had a far worse mission. He had to inform the staff and the larger community what had happened.

Jallah told Princess and the other health workers. News of the young girl who was no longer cured of Ebola spread around town and to neighboring villages. Despite assurance that she was not contagious during the period when she was back home, the local people did not know what to believe. It seemed they were always being told one thing by foreigners, but then seeing something else. Three days later, Pauline died.

Outreach teams had been working around the clock to educate people about the virus and dispel conspiracy theories about Ebola's origin. Now, Pauline's case presented the opposite of what many had been told. Part of the community reverted into denial and suspicion. "That was the biggest challenge we ever had," Jallah said.

Viruses enter environments with no regard for their geographical or political history, their religions, traditions, or cultures. Pauline's case is a testament to how unpredictable epidemics are. Epidemics always progress unevenly, with gains and setbacks. The impulse to keep Pauline's recurrence secret may have been intended to shield an apprehensive community, but the news would have done more harm if it had leaked out rather than being openly acknowledged.

Pauline's case personifies the complexities of outbreaks in real time. Not much was known then about the virus and how it could appear and disappear, spare some and kill others. The unpredictability made it impossible to provide definitive answers. Interagency communication and structural problems furthered the divide between questions and answers. So, too, did limited engagement with the community and lack of transparency, especially in situations where quick decisions had to be made. Outbreak culture and the politics surrounding the virus exacerbated an already turbulent setting.

It wasn't just in the rural areas of Liberia where the volatile environment led to denial of Ebola's existence and distrust of authorities and responders. These struggles also played out in the urban heart of the country. Two hundred and forty miles from Foya, a rash of cases appeared in West Point, one of Liberia's most populous slums, located in the country's capital, Monrovia. Bodies of the dead were left out in the street for days as the city grappled with the growing number of Ebola cases and limited healthcare capabilities. In August 2014, the Ministry of Health converted a West Point school—which had been closed because of the epidemic—into a community care center where Ebola patients could be isolated while waiting for treatment at the hospital. Hundreds of people protested outside. Some claimed that the epidemic was a hoax, and others were angry that patients from around the city were being brought into their community. One night in August, a group of young men broke into the building to steal supplies, and more than a dozen Ebola patients fled into the city. Days later, the Liberian government set up a military perimeter around the neighborhood and prohibited anyone from entering or leaving.[15]

Quarantining, a common containment practice, was used during the outbreak as a means to isolate individuals with Ebola and inhibit the virus's spread. But quarantining an entire community only contributed to the massive death toll. No one besides a designated health worker was allowed to enter or leave the slum. For days, tens of thousands of people were trapped without food or running water. The unsanitary measures left those living under quarantine at risk for exposure to diseases other than Ebola. Months later, when Liberia was declared Ebola-free, President Sirleaf acknowledged that the forced quarantine of West Point had caused more damage than good, especially in fueling doubts about Ebola's existence and suspicions of conspiracy on the part of the government.

Most published accounts of the Ebola epidemic center on a single story of shortsighted, ignorant, and defiant people who refused to alter their traditions even at the risk of death. Simplifying the narrative of the outbreak in this way follows a standard pattern in the history of Western attitudes toward Africa. Nigerian novelist Chimamanda Ngozi Adichie calls out the West for having "a tradition of telling stories of sub-Saharan Africa as a place of negatives, of difference and darkness."[16] Framing the epidemic as having been spread by cultural practices prevents us from understanding the complexity of the threat within the environment, as well as from working collaboratively with people who are at risk. Public health messaging campaigns, for example, focused on communicating critical information about Ebola and correcting misinformation, yet initially the people who designed them paid little attention to the historical, political, economic, and social contexts in which they were delivered.[17] One sign read, "Ebola is caused by a virus. Ebola is not caused by a curse or by witchcraft."[18] Both local and foreign agencies created outreach teams that launched communication campaigns to educate West Africans about the virus. Local people were told that "traditions kill" and were urged to refrain from centuries-old funeral traditions, which included washing the body before burial.[19] Women, the pillars of caregiving in the household, were told to forgo their responsibilities and instead to take their sick family members to be cared for in a hospital. People were dissuaded from physical contact. Some campaigns urged people not to eat bushmeat—meat from wild animals, such as monkeys and bats—because it was a possible source of Ebola virus, and eventually governments across the region temporarily banned its consumption. But bushmeat was an affordable form of protein that had long been eaten with no ill effects.

In some places, control efforts went beyond public health messages to physical enforcement, including curfews, mass cremations, and quarantines. A review of local engagement strategies during the Ebola outbreak concluded that many of the approaches ignored social structures and institutions and involved "scant consideration of local perspectives." Directives were initially delivered in a standardized manner across multiple communities instead of being developed with input from local community leaders, who had a greater ability to change behaviors. According to the researchers, these strategies were "justified as necessary in order to deal with unruly populations who have ignored advice and resisted medical teams due, it is said, to ignorance of biomedicine, irrational distrust and refusal to abandon 'traditional culture.'" Community resistance, however, was not solely rooted in refusal to believe in the reality of the disease or to cooperate with international responders, but had to do with a long history of exploitation and distrust.[20]

Many people were understandably skeptical that medical intervention was beneficial when they saw so many deaths occurring under medical care. Consider, too, that a person with undiagnosed malaria—a common disease in the region with symptoms similar to those of early Ebola—had a higher chance of dying from Ebola if they went to an ETU than if they stayed home.[21] Once coming down with a likely case of Ebola, knowing that the disease was incurable, a person would be faced with the decision of either dying at home or in an ETU. What might appear to be an irrational choice in the midst of chaos and confusion could be reframed to be viewed as the natural decision in a culture that values family involvement and compassionate care for the dying—the only form of care many had ever known.

Another shortcoming of the narrative of ignorance is that it overlooks the many highly educated Africans who were on the front lines of the Ebola outbreak—people like Dr. Khan and Abraham Borbor. Borbor was one of 50 doctors in Liberia. When he wasn't caring for patients at the John F. Kennedy Medical Center in Monrovia, he was teaching the interns and residents who would become the next generation of health workers. Borbor contracted Ebola and was one of the first to receive the experimental drug ZMapp. He initially seemed to respond well to the treatment, but he ultimately succumbed to the disease on August 25, 2014.[22]

Olivet Buck was a high school science teacher turned physician who worked as the medical superintendent of Lumley Government Hospital in Freetown, Sierra Leone. She was one of only two doctors at the hospital. Buck refused to turn away Ebola patients despite overcrowding in the hospital, a lack of health workers, and a shortage of protective gear. Buck probably contracted Ebola because she continued to treat patients without wearing adequate protection. The Sierra Leone government asked that Buck be evacuated to Germany for treatment, but the WHO denied the request. She died on September 19, 2014.[23]

Ameyo Adadevoh, an endocrinologist at First Consultants Medical Center in Lagos, Nigeria, contracted Ebola from a patient who arrived at the clinic after falling ill at the airport in Lagos on July 20, 2014. The patient, a man who was traveling from Liberia, was initially diagnosed with malaria and insisted on being discharged to attend a conference that he had been on his way to.[24] But Adadevoh ordered that he be tested for Ebola and remain at the clinic until the results came in. The man tested positive for Ebola and died four days after admission. Adadevoh died of Ebola a few weeks later. The number of confirmed Ebola cases in Nigeria was contained to 19, nine of whom were doctors and nurses at the hospital where the first patient had been treated. Seven people in total died.[25] Two months after Adadevoh's death, Nigeria was declared Ebola-free. Had Adadevoh not kept the man at the clinic, the epidemic almost certainly would have spread more widely in Nigeria. Dr. Thomas Frieden, then director of the CDC, called the scenario the "scariest moment" in the Ebola epidemic.[26] Nigerians credit Adadevoh as "the woman who saved a nation."[27]

Local responders finally received some recognition in the media when *Time* magazine chose "Ebola fighters" as its Person of the Year. Among the doctors, nurses, and caregivers featured were a number of Africans.[28] But there are at least six hundred more narratives of local heroes similar to these.

Ordinary citizens who worked to overcome their fear and the stigma associated with Ebola also remain relatively underreported. Organizations across the region, including the Liberian NGO NAYMOTE, which had been working in community organizing and civic engagement long before the outbreak, were instrumental in unifying citizens and establishing a sense of local ownership of the Ebola response. NAYMOTE, for example, was able to rapidly coordinate leadership forums addressing health promotion in 60 communities across 13 of Liberia's 15 counties, in part because of its established roots. Local people also felt more comfortable expressing their concerns in forums among people they knew than to government workers or foreign responders.[29]

Some Western media reports centered on defiant villagers who did not understand the Ebola virus. While there was some truth to this story, it represented the epidemic, and the people who were experiencing it, in a simplistic way, focusing on local people as ignorant of the disease's enormity. That kind of simple narrative, in Adichie's words, "flattens [the] experience" and doesn't tell the whole story.[30]

Communicating the complexities of the environment, though, was not easy. Journalists in West Africa grappled with a shortage of credible information in the midst of the epidemic. A survey of African journalists published by the World Federation of Science Journalists found that one of the major challenges in accessing credible information involved the reluctance of government officials to communicate with journalists. Many healthcare workers also reportedly would not speak to journalists, citing the need for government authorization even if they were not subject to restrictions. Journalists reported a lack of collaboration by authorities and health professionals, including "withholding, hiding, controlling and manipulating information." They also reported difficulty accessing official documents and distinguishing rumors from fact. An added physical challenge was the difficulty of reaching remote areas.[31]

We now know that the Ebola outbreak was caused by a complex set of factors. Although skepticism on the part of local people played a part, so did compassion and caring. The two primary groups of people who served as caregivers were also the two most likely to become infected: healthcare workers and women. An estimated 815 healthcare workers in Sierra Leone, Liberia, and Guinea were diagnosed with Ebola between January 2014 and March 2015, and an additional 225 were suspected of having the disease. Two-thirds of those infected died.[32] All who cared for Ebola patients knew the horror of the disease. Many described "living in fear and terror" as they grappled with the decision to go to work and take the chance of contracting Ebola, which they then could potentially transmit to their family members when they returned home.[33]

It is not enough for responders to focus on treating the sick. An equally important part of response is healing the deeper wounds in a community, acquiring cultural competency, and combatting stigma. Limited communication from government and foreign agencies to citizens set the wrong tone early on in the epidemic. Many responding organizations did not initially prioritize building relationships with affected communities.[34] With their focus on trying to control the growing number of cases amid a lack of resources, they gave too little attention to delivering messages that resonated within communities. The result of not developing a detailed understanding of people's perceptions and experiences was resentment and resistance toward Ebola response teams on the part of local people. Many survey respondents said that health workers' lack of understanding about local traditions made it much harder to contain the outbreak.

The distrust in a region that was already firmly rooted in skepticism of the government and foreign entities in some cases led to violence. One survey respondent recalled being chased out of a village in Sierra Leone by residents carrying machetes after a team entered to decontaminate the home of an Ebola-stricken person. Another described being dispatched as a community mobilizer to a village in the southern part of Sierra Leone only to have the village chief order the young men of the community to chase the worker out. "I went there to deliver lifesaving messages about Ebola's spread, transmission and prevention," the participant wrote. "I ran over 3 miles for my life. They never allowed me to speak to anybody."

Even many of those within the community who were part of the response efforts were met with resistance. Some who worked as contact tracers, monitoring the symptoms of anyone who might have come in close contact with someone with Ebola, were considered "betrayals in their communities," since they reported cases of the sick and dead back to the local government, one respondent wrote.

Outbreak responders in Guinea experienced more resistance than workers in the other countries. The International Committee of the Red Cross reported that aid workers in Guinea were subjected to an average of 10 attacks every month during the outbreak.[35] In September 2014, eight local officials, healthcare workers, and journalists attempting to raise awareness of the disease were murdered by a group of angry residents in the village of Womey. Two months later, nearly two hundred miles away in Kissidogou, thieves robbed a courier taxi carrying blood samples being taken to a testing center. Despite public appeals, the cooler bag, which may have contained infected blood, was never recovered.[36] Guinean Red Cross volunteers were also reportedly beaten and stoned while trying to conduct safe burials.[37]

The high level of hostility in Guinea probably had to do with the fact that Ebola was rarely visible to the public, according to Jerome Mouton, MSF's head of mission in Guinea. "It was not like in Monrovia [Liberia], where there were patients everywhere and nobody could deny its existence," said Mouton. "More than 90 percent of the population in Guinea never saw a person who had Ebola."[38]

The number of Ebola cases reported each week in Guinea had been steadily declining by the time Mouton arrived in June 2014. Fewer patients were appearing in the southeast region of the country, where the outbreak was centered. Early CDC and WHO responders who had been in Guinea's capital, Conakry, had not seen an Ebola patient in nearly two weeks. Dr. Pierre Rollin, the CDC's team leader in Guinea and the agency's top Ebola expert, saw this decline as a sign that the outbreak was under control and decided to return home.[39] "Ebola was unexpected," said Mouton, "and we did not consider informing the population and convincing them of the severity of Ebola."

Mouton, who spent the majority of his two years in Conakry, said the initial messages were delivered to an already distrustful community with a top-down approach that disregarded public concerns, further elevating citizens' panic and mistrust. "Our [MSF's] main communication at the time was public calls for help, [that] we needed other actors to step in and help us," Mouton said. "On the population side, that message perpetuated fear."

Response teams were hampered by two main factors: rapid spread of the deadly virus, and rapid spread of misinformation. Workers who took away infected people and dead bodies had to work quickly, with little time to explain what they were doing and why. That's one of many reasons conspiracy theories took hold, Mouton said. People did not respond well to being told what to do (and not to do) when they did not understand the full scope of the problem or trust the sincerity of the response teams. "With the little resources we had in the beginning, it was impossible to expect people to act rationally," he said. "We were not in a position where we could invest in resources to communicate to the community via a public service announcement."

It is inaccurate to ascribe people's resistance solely to their traditions or supposed lack of modernity, he added. An additional factor had to do with how they accessed healthcare. "If you are people who have an accessible health care system, the way you seek help will be different," he said. People in developed countries whose healthcare is free or covered by insurance are likely to seek out health services frequently, even for minor conditions. But in places like Guinea, where health systems are not adequate or accessible to begin with, it is unreasonable to expect that people would seek out conventional medical care.

Ebola treatment centers became seen as equivalent to slaughterhouses. Patients entering ETUs were isolated from everyone except healthcare workers wearing protective gear and masks that hid their faces. Strangers were doing the job that was traditionally fulfilled by family members. The isolation was traumatizing for patients and for their relatives, who realized they would never again be able to interact with the patient or attend them in death.

A successful response would have meant involving local people in containment efforts from the very beginning. Mobilizing communities requires taking time to build trust and a strong local infrastructure before a crisis, rather than trying to pull these elements together as an outbreak emerges.

Mouton said that by the end of June 2014—within a month of his arrival—village leaders had denied MSF access to 28 villages known to have Ebola cases. "Even though they were dying without us, they still didn't want us coming in," he said. "Without communication efforts, it was hard to fight the epidemic."

Getting the words right was important, but even more vital were the nonverbal messages. Mouton decided that if Guineans were to understand the disease and efforts to protect them they had to see the public health efforts for themselves. Facing resistance after less than a month of arriving in Guinea, Mouton decided to implement a strategy of transparency. He allowed free access to newly built treatment centers before they opened. He approached community leaders and invited them to visit the three centers that were being built—one in Guéckédou, another in Conakry, and a third in Macenta. "We made open fences so people could see the buildings," Mouton said. "They weren't hidden behind tall walls." In addition, they urged survivors to talk about their experiences "so they could affirm, 'no, they're not killing us.'"

Mouton found that being transparent and involving people in the response efforts helped them understand what was happening within their community and how they could be a part of the solution. It was more complicated and time consuming than putting out public service announcements, but it warded off rumors and misinformation. "Once people were convinced of the value we and these centers had for their communities, and they could be a part of the effort, our work became much easier," Mouton said.

Although these efforts did help to build trust in the community, not everyone was convinced. To this day, many Guineans question the reality of the Ebola virus. On one of his final days in Guinea, Mouton, who had been driving a car marked with the letters M-S-F, pulled over onto a side street in Conakry. Two Guinean men stood together in front of a store, their eyes fixed on him as he stepped out of the car. One of the men nodded.

"Ebola doesn't exist," the man said, loud enough for Mouton to hear.

The second man squinted in disbelief and nudged his friend's arm with the back of his hand.

"Come on man, Ebola does exist," the second man told his friend.

"That sums up the situation to this day," Mouton recalled. "I think it's because he experienced the confusion but never saw what we saw."

Hundreds, perhaps even thousands of Ebola cases and deaths could have been prevented through a strategy of empathy, which requires understanding the beliefs of the people whose behaviors responders are working to change. In the regions where rumors were most rife, transparency early on might have altered the course of the epidemic. Such an approach is key to countering suspicion and misinformation. Most important, it is vital to preventing epidemics of fear.

"Fear has never helped people think or act rationally during an outbreak, especially when death is at stake," Mouton said. "When we are scared, we no longer act sensible."

6

EPIDEMIC OF FEAR

The plague was nothing; fear of the plague was much more formidable.

—Henri Poincaré, *The Value of Science*

BOSTON UNIVERSITY infectious disease physician Nahid Bhadelia stepped off the plane at Logan Airport in Boston, Massachusetts, and rolled her carry-on bag to customs. It was November 2014, and Bhadelia had just flown in from Sierra Leone. She was returning from her second visit to Port Loko to care for patients with Ebola.

One month before, on October 8, the first patient diagnosed with Ebola in the United States had died, resulting in massive hysteria and fear that the outbreak had made its way to the developed world. The CDC issued new screening requirements for anyone traveling into the United States from Guinea, Liberia, or Sierra Leone. Travelers had to answer a set of questions when they left Africa and then again when they arrived in the United States. They were then required to report daily health information to state and local health departments for 21 days, the maximum time for someone infected with Ebola to show symptoms.[1]

A customs agent picked Bhadelia out for additional screening. She followed the agent to a small, colorless room. Two plastic chairs were separated by an empty table. A phone hung on the wall. Bhadelia and the agent sat down.[2]

"Have you been to West Africa?" the agent asked, looking at a sheet of paper containing a list of questions.

"Yes, I have," she replied.

"Have you been in an area where Ebola is active?" he asked.

"Yes, I have," she replied.

"Have you taken care of Ebola patients?" he asked.

"Yes, I have," she replied.

The agent lifted his eyes, raised his head, and placed his pen on the table.

"No one has said 'yes' so far," he said. He walked over to the phone on the wall and dialed out.

"Yes, sir, the doctor said 'yes,'" the agent told the person on the other end. "What happens next?"

Minutes later, a nurse working for the CDC walked into the room dressed in full PPE. The agent beside her looked exposed, compared with the fully protected nurse. Bhadelia had also just spent 12 hours among hundreds of other airline passengers and airport workers. The nurse was the only one wearing a protective suit.

"I'm going to take your temperature," the nurse said, her voice muffled through the plastic face mask. She rummaged through a bag of medical equipment, looking for a disposable thermometer. Her search came up empty.

"Here, I have one," Bhadelia said, unzipping her carry-on bag.

The nurse gave Bhadelia a dubious look.

"Yeah, we're not going to let you use your thermometer," she replied.

Bhadelia showed no signs of Ebola: no fever, no aches or pains. Still, the CDC recommended that she quarantine herself in her apartment for two weeks and check in to the state's public health department every day with a temperature reading. This quarantine protocol for travelers returning from West Africa was relatively new, and while it appeared to ease some public concerns, clinicians like Bhadelia believed that active monitoring for symptoms, not isolation, was a more adequate measure, as well as being more feasible and just as effective.

At the height of the outbreak in July, there were no restrictions on travelers. Researchers Nathan Yozwiak and Stephen Gire had returned to Boston from Sierra Leone just days before Sheik Humarr Khan is suspected to have contracted Ebola. The two Americans did not undergo a special screening process. They were never monitored, nor did they have to send reports to any health departments. The Sabeti Lab, having worked for many years on another deadly viral disease, Lassa fever, had established its own monitoring and quarantine practices, in partnership with the travel clinic of Harvard University and Massachusetts General Hospital. Yozwiak and Gire maintained communication with physicians at the clinic and agreed to immediately seek care if they experienced a fever within the 21-day window following their return.

Neither ended up getting Ebola. Other groups working in the region, however, had not set up formal procedures, and any policies that were in place had not yet been established on a national level. At that point in the outbreak, "there was a lot of 'fly-by-the-seat-of-your-pants' stuff that was going on," Yozwiak said.[3]

Bhadelia followed the recommended protocol. When her apartment building manager found out that she had returned from the front lines of an Ebola-stricken country, the management company returned her month's rent and requested she quarantine elsewhere. Bhadelia left and stayed with her parents nearby. The company's response was benign compared with what some other health workers faced when they returned from West Africa. Debbie Wilson, an American nurse who was volunteering with MSF in Liberia, described returning home to find neighbors petitioning for her eviction and crossing to the opposite side of the street when they saw her coming. Some health workers returning from West Africa actually were evicted. Others were refused entry into their home states. Some had their children disinvited from birthday parties.

The stigma was not limited to individuals returning from West Africa. Staff members at the University of Nebraska Medical Center, one of the hospitals that was designated as a receiving center for Americans returning with Ebola, received death threats daily. Nurses at Bellevue Hospital Center in New York City, where one of the American Ebola patients, Craig Spencer, was treated, were refused service in restaurants and hair salons.[4]

The similarity between West Africa and the United States was that people in both places did not have a good understanding of the nature of the disease, how it spread, or what their risk was. Inaccurate perceptions sometimes got in the way of an effective and reasonable response. However, West Africa was facing a true disease epidemic, whereas in the United States, the only epidemic was fear.

Unlike the situation in West Africa, where thousands were infected with Ebola, only eleven people were treated for Ebola in the United States during the epidemic, two of whom died. Seven of the eleven became ill while in West Africa, and one of them died, despite being evacuated and treated in the United States.[5] The only two cases of Ebola that were transmitted within the United States came from Thomas Eric Duncan, a Liberian man who was the first person to develop Ebola symptoms after arriving in the United States. Duncan was cared for at a Dallas hospital, where two of his nurses became infected. These confirmed cases set off fears that the United States would experience an outbreak similar in proportion to what was occurring in West Africa.[6] Three days before Duncan died, government health officials in Sierra Leone reported

121 Ebola deaths in a single day—a number that made some headlines, though it went largely unnoticed among Americans. Meanwhile, given the tiny number of cases in the United States, the chance of an American contracting Ebola was close to zero. The magnitude of the outbreak "over there" seemed insignificant compared to what was happening "over here." Duncan's case resonated more strongly with many Americans because it was closer to home.

The resulting panic had wide repercussions. Commercial airlines and air freight companies stopped flying to West Africa, severely limiting aid to the most vulnerable parts of the region. American politicians with no knowledge of the disease called for quarantines and limits on immigration to quell their constituents' fears. Scores of people who had recently returned from West Africa with "flu-like symptoms"—a vague complaint that could possibly signal infection with Ebola—trickled in to hospitals across the country, setting off a media firestorm of hypotheticals: What if the patient has Ebola? Where did it come from? Who else could be infected? What if it spreads throughout the state?

Academic medical centers, some of which had discouraged their own employees from volunteering abroad, found themselves having to train staff and build up capacity to care for Ebola patients. Some staff members at smaller community hospitals took personal days or sick time off, mainly because they knew that if a person with Ebola came to the hospital, they wouldn't have the slightest idea what to do.

As Bhadelia was serving out her quarantine, dozens of clinicians received a call to report to Massachusetts General Hospital in Boston. They were met by frenzied news reporters and cameras staked out in front of the hospital. Rumors were flying that a patient arriving at the hospital might have Ebola.

Dr. David Hooper, chief of the Massachusetts General Hospital's Infection Control Unit, had initiated an emergency telephone tree to alert medical staff, as part of the protocol put in place by the CDC.[7] The patient in question had recently returned from working in Liberia. The Boston Public Health Commission, which had been monitoring his health, decided that his persistent fever and chills matched the symptoms of Ebola. The man was not a medical worker and had not come in contact with Ebola patients during his trip. But on a frigid Tuesday morning in December, dozens of staffers at Massachusetts General Hospital donned full PPE to receive him.

In the coming days, hundreds of hospital employees, from physicians to food deliverers to security guards, were involved in the patient's care. He was first placed in a special holding room in the emergency department. Within 24 hours of his arrival, three walls had been erected to serve as a specialized isolation room. Later, an entire section of the intensive care unit was cleared out to house him.

Outside the hospital walls, media mania continued. Photographers, reporters, and producers huddled on a street corner waiting for updates. But once the news organizations found out the patient had tested negative for Ebola, they turned off the cameras, put down their pens, and moved on to the next story. The patient was treated for malaria, and less than a week after he had been admitted, his symptoms cleared and he was sent home. It was the Ebola case that wasn't. Other hospitals across the country were facing similar false alarms. Social media and the internet have made it possible for rumors and panic to spread faster than ever.

The national panic over Ebola bore some resemblance to the early days of the AIDS epidemic in the 1980s. The stigmatizing of American healthcare workers and African immigrants, the amplification of hysteria by politicians and the media, the shunning of entire groups of people based on fear and suspicion, and calls for extreme quarantining of suspected cases are all common themes.[8]

While AIDS was indeed an epidemic in the United States—unlike Ebola—the social and political frenzy in both cases epitomized the worst of outbreak culture. The viruses receded into the background while policies were steered by emotion rather than evidence.

Media framing, too, exploited people's fears. Ebola hysteria in America was due, in part, to accounts that focused on perceived threats from an "other" and that tended to exoticize African culture. Media portrayal of the epidemic and of Africans not only mirrored society's perceptions, but simultaneously perpetuated the fear. On August 21, 2014, *Newsweek* published a cover story on Ebola which featured a picture of a chimpanzee and suggested that bushmeat illegally imported from Africa could spark an Ebola outbreak in the United States.[9] The story was headlined "Smuggled Bushmeat Is Ebola's Back Door to America." A graphic titled "Don't Eat That Chimp!" listed some of the most popular animals consumed as bushmeat among Africans and explained that chimpanzees and monkeys might have been the original source of the HIV virus. Some bushmeat was available in the US black market, despite USDA regulations that prohibited importation of meat products from any African country, but there was only limited evidence to suggest that bushmeat might be a culprit in the transmission of Ebola. The headline, in particular, seemed to be stoking fear based on speculative theories. With the exception of the first suspected transmission of Ebola from bat to human, it remains highly unlikely that any of the more than 28,000 secondary cases of Ebola were transmitted by anything other than human to human contact. *Newsweek*'s chimpanzee cover became yet another misrepresentation of the epidemic. It furthered the perception that Ebola is an exotic illness transmitted by "others" whose strange, barbaric practices could be the death of us.

African immigrants, in particular, were stigmatized in the same way as communities suffering from AIDS were stigmatized in the

1980s. Sensationalized news reports fueled concerns that Africans would spread the disease. Problems arose especially in cities with high Liberian and Nigerian immigrant populations, such as New York City, where there were numerous reported cases of bullying at school and work when suspected and confirmed cases began to emerge in the United States. Many Africans felt they had to hide their ethnicity to avoid public shaming.[10]

Television networks' obsession with Ebola also heightened public concern. Evening broadcast and cable news shows on ABC, NBC, CBS, MSNBC, Fox News, and CNN collectively aired nearly one thousand segments about Ebola between mid-October and early November 2014.[11] Coverage focused mostly on domestic cases, naming individual patients and following their stories, which made sense because of the small numbers of infected people in the United States, but which also intensified public concern.[12] In a cycle of public interest fueling media coverage, and media coverage in turn fueling public interest, there was constant demand for new reportable information on cases, prevention, and response, which led to further public outrage that not enough was being done to combat the outbreak.

Politicians stepped in, conveniently during midterm election season, to mandate policies that health experts knew did nothing but provide a false sense of security. Directives imposed on health-care workers returning to the United States after treating Ebola patients abroad—like those imposed on Bhadelia and nurse Debbie Wilson—were based on the public's fears rather than on sound medical judgment, making the job of public health experts and clinicians caring for suspected cases more challenging.

The way quarantine was carried out, for example, was an antiquated approach, yet the concept became an obsession in public discourse. Quarantine, which differs from isolation, is a "coercive social-distancing model that removes those who have been exposed to disease and *might become sick*, but who have not tested positive for disease or manifested clinical symptoms of disease, from the community." Isolation "removes those who *are confirmed sick* from the population for treatment and recovery."[13] While quarantine can be effective if used appropriately, it was unnecessarily applied in the United States during the Ebola outbreak. Ebola is not contagious until a person is symptomatic. Since symptoms appear gradually—usually beginning with a high fever, muscle aches, and abdominal pain—a person has enough time to seek medical care before they either become capable of spreading the disease to others through fluid secretion or become incapacitated. In a country with a developed health system like the United States, individuals can be easily tracked and connected to medical care for their symptoms. Even in West Africa, with its inadequate health systems, quarantine as practiced was ineffective, and government mandates were counterproductive and damaging. Quarantine prevented the free movement of necessary medical supplies and personnel to areas in need.[14] In Liberia's West Point neighborhood, for example, quarantine looked more like entrapment. It led to increasing health problems because of unsanitary living conditions and worsening distrust of responders.

Still, quarantining became the public's preferred solution to preventing an Ebola outbreak in the United States. The power to enact quarantines resides with state governments, so the decision does not require national coordination or medical leadership. In the few states that mandated quarantines, governors implemented these policies through health departments without judicial or scientific input.[15] This suggests that the practice may have served a political purpose, rather than one of public health.

Thomas Duncan arrived at Texas Health Presbyterian Hospital in Dallas on September 26, 2014, complaining of abdominal pain, dizziness, nausea, and a headache. The hospital had implemented a policy a month earlier requiring a travel history from each patient, but Duncan had not filled one out, and so health workers were unaware that he had recently returned from Liberia. Despite having a 103-degree fever during his 4-hour hospital stay, Duncan was diagnosed with "sinusitis and abdominal pain" and sent home. Less than a week later, he returned to the hospital. It was then that he was diagnosed with Ebola.

What happened within the hospital walls after that remains unclear. The CDC contends that its team of experts arrived in Dallas within hours of Duncan's diagnosis. An independent review of the hospital's response, however, found that agency personnel did not arrive until three days after Duncan's hospital admission. The review also suggests the CDC did not adequately prepare the hospital staff in the use of PPE and other procedures to avoid contamination.[16] Soon after Duncan's case, the CDC made changes to its recommendation on PPE.[17]

Meanwhile, misinformation about Ebola was circulating widely, and Duncan's case elevated the level of fear. The CDC, attempting to ease concern, assured the public that US hospitals could "safely manage" Ebola patients.[18] But on October 8, Duncan died—the first person to die from Ebola in the United States. Then, two nurses who had cared for him became ill. The CDC was widely criticized for projecting overconfidence in the ability of hospitals to handle Ebola.[19] Barclay Berdan, CEO of Texas Health Resources—which owns the hospital where Duncan died—later criticized the CDC, saying that its role in Duncan's case was unclear and that CDC trainings and guidelines were insufficient.[20] The situation heightened public distrust of government health officials.

Despite CDC-provided trainings in some hospitals and the dissemination of a preparedness checklist, the majority of conventional medical centers had nowhere near the level of readiness required to manage multiple Ebola patients, or even just a single one.[21] Only four hospitals had specialized biocontainment facilities, and in most hospitals, only a very few staff members were adequately trained in procedures relating to dress, clinical care, and waste disposal, which needed to be meticulously followed to avoid contamination. After Duncan's death, many hospitals across the nation increased capacity and training to adequately monitor, care for, and treat Ebola patients. The CDC designated a total of 55 hospitals across the nation as Ebola treatment centers.[22]

Duncan's death also exposed the absurdities of improper implementation of quarantines. Duncan's fiancée, Louise Troh, their son, and two nephews were forced to remain in the family apartment for several days, even though it contained soiled linens that may have been contaminated. Authorities had trouble finding a company with a permit to transport biohazardous waste. Keeping the family, who showed no symptoms, in a potentially tainted space served only to raise their risk of becoming infected—a scenario that also occurred in parts of West Africa. Fortunately, Duncan's family members did not contract Ebola.[23]

The CDC faced a dilemma as the public turned to health authorities for answers. Communicating uncertainty can undermine confidence in expertise, yet in unpredictable or fluid situations, attempting to make predictions or provide answers that might end up being wrong can lead to a loss of credibility.[24]

The CDC stated that Duncan's nurses probably became infected because of a protocol breach at the hospital. But the public soon learned that there was no Ebola protocol at the hospital. They also learned that the CDC had permitted one of the nurses to fly from Dallas to Cleveland, Ohio. Even though the CDC acknowledged that it should have sent out an infection team to the hospital and promised to implement nationwide hospital trainings, the damage was done. The public's perception of experts' inadequate understanding and preparedness increased fear and devastated public trust.

One of the two infected nurses, Amber Vinson, flew to Ohio a few days after Duncan's death, before she was diagnosed with Ebola. She had called the CDC multiple times before flying to report that she had a fever of 99.5 degrees. Since her temperature fell below the agency's threshold of 100.4 degrees, she was not considered "high risk" and was cleared to fly.[25]

While in Ohio, Vinson tried on wedding dresses at a bridal store. Once she was diagnosed with Ebola, Ohio state health authorities mandated that the bridal shop immediately close. Area schools suspended classes. The shop was never able to reopen because the public continued to have unfounded fears that the merchandise was contaminated with the Ebola virus.

Eighteen Ohio nurses who were on the flight with Vinson from Dallas to Cleveland, returning from a conference, were placed on paid leave from the hospitals where they worked. At the time, "there was no information available about where individuals were seated on the plane, and whether they would have had direct contact with [Vinson]," said Dr. Jennifer Hanrahan, medical director of infection control at MetroHealth Medical Center, where eight of the nurses worked.[26] A flight seating chart listed passengers' assigned seats, but the hospitals could not verify whether people actually sat in their assigned seats. "The decision to place them on leave had to be made without knowing which nurse was in close proximity to Vinson," said Hanrahan.[27]

The Ohio Department of Health was alerted of the nurses' leave, but according to Hanrahan, patient privacy laws kept the hospitals from providing detailed information. Hanrahan contacted the CDC directly for guidance on when the nurses would be allowed to return, even though a state's public health department would typically be the primary connection to the agency. Although there is no centralized CDC number to call when discussing a potential public health threat, Hanrahan expected that contacts at the agency would at least be aware of the public scare. The person Hanrahan spoke with had no idea about the situation unfolding in Ohio, which led to more confusion. "There was a lack of communication between agencies who had information and those of us who needed it," Hanrahan said.[28]

In addition to being placed on leave, the nurses were also urged to remain in their homes for 21 days, the maximum amount of time to the onset of symptoms. This quarantine protocol was loosely placed. The nurses were advised to avoid leaving their homes, but they were not told to stay away from direct contact with family members who lived in the home. "We did not give strict rules about what they were supposed to be doing," Hanrahan said, adding that in part, it was because the medical staff knew that the nurses' risk for Ebola was low.[29]

The hospitals, however, put up a united front and decided to place the nurses on leave under an "abundance of caution." Even after learning that the nurses had not been sitting near Vinson and had not had any direct contact with her during the flight, they maintained the nurses' leave for no other reason than to avoid public backlash. "The decision had already been made and we couldn't turn back with what we made public," Hanrahan said, adding that although the quarantine did not protect anyone, people thought it did.

True quarantining would involve isolating an individual from the rest of the population in a single location for 21 days. The person would have to be supplied with food and other needs while being regularly monitored. To this day, there are no facilities in the United States prepared to implement the textbook definition of quarantine or to isolate more than one symptomatic person at a time. Even MetroHealth Medical Center, which received designation as an Ebola treatment center after Ohio's Ebola scare, does not have this capability.[30]

Two public health measures are generally used for controlling the spread of an infectious disease when there are no effective vaccines or treatment. The first is isolating the person who is symptomatic or is confirmed to have the disease. The second is tracing and quarantining anyone who has come in contact with the infected person.[31] Both of these measures must be applied quickly to prevent spread of the disease. However, the efficacy of isolation and quarantine depends on the nature of the pathogen. Quarantining may be a more appropriate option for airborne diseases, since it is possible to contract them simply by being in close proximity to an infected person. However, that too depends on the incubation period of the disease and the type of exposure. With SARS, for example, there are a few days after the onset of symptoms during which a person can be isolated to effectively interrupt transmission.

Other communicable diseases, like Ebola and AIDS, are much more difficult to contract. Both the HIV and Ebola viruses are transmitted through fluids, not through the air, and require direct contact. HIV, however, is transmittable even before an infected person shows symptoms of the disease, thus making it much more difficult to isolate people swiftly and trace everyone they have been in contact with. In contrast, there is no evidence that Ebola is contagious until a person is symptomatic, thus making it difficult to justify quarantining asymptomatic individuals who were only in the presence of an infected person.

Since days can go by after infection until a person shows signs of Ebola, close and frequent monitoring of symptoms is much more likely to curb transmission than quarantine. Yet, for a community in fear, monitoring is likely to be perceived as too passive, compared with the more active approach of isolation and confinement.

A major practical and logistical challenge of quarantines is compliance. During the 2002–2003 SARS pandemic, residents of an apartment complex in Hong Kong fled before officials came to move them to a quarantine facility.[32] After SARS spread to Toronto, health authorities quarantined 23,000 people, but only 57 percent complied.[33] Lack of compliance also occurred in affected West African cities, towns, and villages during the Ebola outbreak. Some people escaped by bribing military or police officers standing guard at the perimeter of sealed-off zones, undercutting the quarantine efforts and highlighting the futility of forced policies in a community accustomed to bending the law. In these cases, quarantining became a policing effort led by law enforcement. Politicians and the fearful public were more likely to advocate for it than physicians. It offered a show of action, when more effective alternatives like education and monitoring seemed too weak.

The attempt to forcibly impose quarantine was taken to an extreme in late October 2014, when a Maine nurse, Kaci Hickox, was isolated for 80 hours immediately after landing at Newark Liberty International Airport in New Jersey. Hickox was returning from Sierra Leone after volunteering with MSF to treat Ebola patients. Despite never having displayed any symptoms of Ebola, Hickox was mandated by the state of New Jersey under then governor Chris Christie to stay in an isolation tent near the airport. She was outraged by the experience, and stark images of her appearing trapped surfaced in the news. After she was released, Maine governor Paul LePage and state public health officials attempted to quarantine her in her home, even though she tested negative for Ebola twice. Ultimately, they failed to secure a court order to force her confinement, and Hickox defiantly appeared outside on many occasions, riding her bike through her neighborhood. Hickox sued

Christie, claiming false imprisonment and invasion of privacy. The two parties eventually reached a settlement agreement, which included establishing a patient's "bill of rights" to become part of the state's quarantine policies.[34]

Although quarantines can be effective in stopping the spread of a pathogen, when they are imposed punitively, they can have an adverse effect. The examples of the Duncan family and Kaci Hickox were some of the most heavy-handed in the United States during the Ebola outbreak, but a similar draconian approach to quarantining has been seen many times before. It often backfires by discouraging individuals from wanting to participate in containment efforts and helping to exacerbate hysteria and fear. If, alternatively, quarantine was shown to be a caring and comfortable experience, fewer individuals would resist. As an extreme counterexample, during the HIV epidemic in the 1980s, Cuba opened up sanatoriums that were so pleasant for residents that many continued to live there well after quarantines were lifted.[35] These sanatoriums also drew some ethical concerns, however. Some viewed the facilities, where all inmates were required to participate in an HIV education program, as prisons, and patients were sometimes used as test subjects for experimental treatments. Patients who left after the prescribed eight weeks and engaged in unsafe sex would be permanently quarantined.[36] The sanatoriums eventually shut down, despite resistance from some of the residents. Quarantine must be designed in such a way that individuals choose to follow protocol during the entire monitoring period.

When NBC News chief medical editor Dr. Nancy Snyderman violated a voluntary 21-day Ebola quarantine after a trip to Liberia to cover the epidemic, the public response was uncompromising. Snyderman and her production crew, who had been in contact with a freelance cameraman who contracted Ebola, were asked by the state of New Jersey to quarantine after returning home, in order to

minimize the risk of an outbreak. When Snyderman, who promised to comply, was later spotted inside a car in a New Jersey restaurant parking lot, the state issued a mandatory quarantine. She was monitored daily, with visits by state public health officers. Snyderman recalled one official acknowledging the worthlessness of the quarantine during a visit, saying, "I know it's overkill, but it's fun to come in and see you guys." The official was confident that Snyderman didn't have Ebola and wasn't at risk of spreading the disease. It didn't seem to matter. Snyderman told the *Hollywood Reporter* that her 88-year-old mother was denied the flu shot because the senior center where she lived feared she had been in contact with Snyderman. Wanted posters were put up in her town asking people to contact police if they spotted her outside.[37]

"We were not sensitive to how absolutely frightened Americans were," Snyderman told the *Today* show.[38] Snyderman resigned from NBC in March 2015 as a result of the controversy, after nearly a decade with the network.

Public fears, which were wholly divorced from reality, spread far beyond New Jersey and other states dealing with suspected and confirmed cases. A November 2014 opinion poll found that Americans ranked Ebola as the third most urgent health problem facing the country, just below cost of and access to healthcare. Ebola was ranked higher than any other disease, including cancer and heart disease, which together account for nearly half of all US deaths each year.[39] The public's high ranking of Ebola was due in part to widespread misunderstanding about how the virus spread and disbelief in official pronouncements. A public opinion poll conducted by Harvard University about Ebola in mid-October 2014 found that an estimated 85 percent of Americans believed that someone could get Ebola if he or she was coughed or sneezed on by a symptomatic person. Nearly half of those polled said a person could pass on

the virus before having any symptoms. Both of these statements are unlikely to be true.[40]

Media coverage and politics both played a part in public perception, but so too did fragmented communication among public health agencies, those expert voices intended to ease public concerns. Some information released by agencies and medical experts was later revised or even retracted completely. For example, the CDC continuously worked to update its website with general information about Ebola and how it is transmitted, but updates sometimes seemed to contradict previous information, only furthering public confusion. Only 31 percent of the public said they trusted US public health officials to share complete and accurate information about Ebola. Forty percent said they did not trust information about the Ebola outbreak from one specific agency: the CDC.[41]

In the United States, unreliable information circulated faster than the virus, which could well be considered the lesser of two evils. Still, the Ebola outbreak serves as an example, in both the United States and West Africa, that the journalistic process of responsible reporting on developing information, reliable source gathering, and fact-checking must improve during public health crises.

The epidemic of fear had swept through American public discourse. Meanwhile, behind the public's view, a subtler yet more detrimental scene was playing out among those tasked to protect the affected, treat the infected, and stop the virus from spreading.

INVESTMENT AND ACCOUNTABILITY

Infectious diseases remain one of the biggest risks facing humankind. Few events are capable of equal damage to human lives and livelihoods. Yet the global community spends relatively little to protect populations from the risks of pandemics. Compared with other high profile threats to human and economic security—such as war, terrorism, nuclear disasters, and financial crises—we are underinvested and underprepared.

—Commission on a Global Health Risk Framework for the Future

BY AUGUST 2014, 35 health workers at Kenema Government Hospital had succumbed to Ebola.[1] Dr. Sheik Humarr Khan was among them. Local responders had been vital to beating back the epidemic, yet they were the most stigmatized and least cared for group. Many of them had died, and those who survived had even more patients to take care of. Fear and suspicion among family members and neighbors prevented them from returning to their homes.

Days after Khan died, the nurses at Kenema, racked with grief and anger, went out on strike, refusing to return to the Ebola ward without a paycheck and proper protective gear. The country's Ministry of Health and Sanitation had promised workers the equivalent of $100 per week of hazard pay in return for their commitment to caring for Ebola patients. Workers were used to waiting weeks to get their paychecks, but during the outbreak, those weeks turned into months with no sign of delivery. Demonstrations outside the hospital perpetuated the perception by Sierra Leoneans that the government had orchestrated the epidemic and that hospitals weren't to be trusted. Meanwhile, conditions for patients worsened inside the already strapped Ebola ward.

People in Kenema awoke one morning to find three corpses exposed and abandoned near Kenema Government Hospital. The night before, a few burial workers had broken in to the town's morgue and removed more than a dozen bodies. One body was reportedly left near the hospital manager's office, and two others were laid out at the entrance to the hospital. At least a dozen others were later found on hospital grounds.[2]

For months, local health workers, including burial teams, had watched as foreign agencies spent thousands of dollars to transport and care for their own staff. They also faced repeated confrontations, escalating demonstrations, and stigma in their own neighborhoods, villages, and homes. The workers had reached their breaking point, and this act was symbolic payback for the poor conditions they endured.

Donations had been pouring in from developed countries, yet local responders received few of the benefits. Less than 2 percent of the $3.3 billion raised globally to fight Ebola was set aside for them.[3] The majority of the money went to Western agencies, including more than one hundred nongovernmental organizations, and to the UN.

Boston University infectious disease physician Nahid Bhadelia, who had spent weeks in Kenema in August, had come to know the local nurses. She knew about their families, their fears, and the frustrations of the people they were working so hard to save. She knew the instability the nurses faced when foreign workers dropped in to Kenema for a few weeks and then returned to their epidemic-free countries. After leaving Kenema, she learned that the nurses were still owed months of back pay. She understood how detrimental it was to their morale, and how much it impeded the progress of containing the Ebola epidemic in the entire country. Bhadelia started a GoFundMe campaign to raise funds for the workers. She set a $50,000 goal, which she calculated would help to pay about 70 workers. By August 2015, the campaign, which Bhadelia described as "a moral obligation," had raised about $42,000, including a few thousand dollars of her own. She cashed out the money, packed some into her carry-on bag, and divided the rest up among colleagues who were flying back to Sierra Leone with her. The money was hand delivered to Kenema Government Hospital and divided among 63 of its workers. Bhadelia said the money went for basic livelihood needs—school uniforms for kids, a motorcycle, nursing classes, and care for orphans of nurses who had died from Ebola. "I wasn't trying to change the system," Bhadelia said. "Just satisfy a need."[4]

Members of the Sabeti Lab had also set up an Ebola relief fund in the fall of 2014 to support the orphans of the many fallen health-care workers at Kenema. They had used Stephen Gire's nonprofit organization Congo Medical Relief to collect and distribute the funds.

It wasn't only Kenema workers who were in need. An estimated one-third of international money allocated to outbreak response across the affected countries of West Africa never made it to its intended recipients.[5] Hospital workers across the region weren't getting paid, and, at the community level, education and outreach programs never received the money they were promised.

Findings from our survey suggest that problems with the distribution of financial resources were caused in part by corruption, but that accountability was also a major challenge. Donors waited too long before requiring local governments to report how the money was being spent. "Hoards of money and resources were flowing into the affected countries that did not have the absorptive capacity to make good use of them," one survey respondent wrote. There were no systems in place to handle proper distribution of money or to record where the money went.

Aid money was most commonly wired to banks with clearance from the government. In some cases, payments were made out to individuals instead of organizations. It was often unclear who was authorized to distribute funds. Hospitals and aid organizations faced incredible bureaucratic hurdles when requesting money. Sometimes a request was met with silence, and other times the response was that there was no more money.

Only a few months after the epidemic ended in Liberia, the country's president, Ellen Johnson Sirleaf, launched a crackdown on organizations accused of mishandling $1.8 million in donor funds. Sirleaf ordered police to shut down the Liberian branch of the Red Cross in March 2016, suspended the chapter's top officials, and dissolved its board of directors. The umbrella organization, the International Federation of Red Cross and Red Crescent Societies (IFRC), undertook an audit of Ebola operations in all the affected countries and prepared a report, made public a year later, which found evidence of fraud in Liberia.[6] Unfortunately, by the time the audit was complete, Liberia had already received the majority of its foreign assistance.

The IFRC's investigation revealed that almost $6 million had been misappropriated in Liberia, Guinea, and Sierra Leone. The organization said it was "outraged" after the investigation concluded that the irregularities were due to fraudulent practices between its own staff members and bank and government officials. In Liberia, an estimated $2.7 million was lost because of fraud related to inflated prices. More than $2.1 million went missing in Sierra Leone, probably because of collusion between former IFRC staff and bank employees, and Guinea lost more than $1.1 million due to fake and inflated customs bills.[7]

Although this report dealt with only a small percentage of relief dollars, the remainder of the money took just as convoluted a path from donor to recipient. There is no global financial governance during public health emergencies, experts tell us, which contributes to the dysfunction of outbreak culture. Funding a natural or public health disaster response has never been simple, and the Ebola outbreak was no different. Our survey found that funding was a recurrent challenge for clinical responders, who felt incredible pressure to perform without adequate support. Although inadequate funding affected all who were involved in any capacity during the outbreak, it was especially problematic at the start. One respondent commented on the "delay in readily available funds to respond appropriately during the start of the outbreak, which could have averted the scope of transmission."

An efficient financial response with appropriate oversight would have altered the course of the outbreak in West Africa. The region would never have experienced such a large-scale epidemic if the money had been invested there in advance. Fighting off an outbreak is much costlier than preventing one. On a global scale, the estimated cost of the core capacities needed to prevent, detect, and respond to outbreaks is $3.4 billion annually, an amount significantly less than the $60 billion to $570 billion that is lost annually on pandemics.[8] An established health system would not have stopped the initial spread of Ebola, and might not even have prevented it from developing into an epidemic; however, a stronger health system would have enabled quicker and more comprehensive action to limit the spread and ensure that people were treated with the care and dignity they deserved.

In 2012, before the Ebola outbreak, the governments of Liberia, Sierra Leone, and Guinea collectively spent an estimated $280 million on health services. Bringing that up to the average level for developed countries would have required an additional $1.58 billion.[9] This gap seems large—and indeed it is—but it is a drop in the bucket compared with the amount spent to contain the epidemic. The epidemic also ravaged the economies of all three countries. Liberia, for example, estimated that it would need $1.3 billion just to recover export revenues it lost due to travel and trade restrictions (both formal and not) during the outbreak.[10]

By December 2014, international governments, including the United States, had pledged a total of $4.3 billion to fight Ebola in Liberia, Sierra Leone, and Guinea—15 times the combined annual government health budgets of the three countries. If the money had been used for prevention, it would have been a better investment; the sum was "nearly three times the annual cost of investing in building a universal health service in all three affected countries," according to a 2015 report by Save the Children.[11] But note that although $4.3 billion was pledged, less than half of that amount was actually sent to the affected countries.

Toward the end of the outbreak, the UN held an International Ebola Recovery Conference, where international donors, including the African Development Bank, the European Union, and the US government, pledged $3.4 billion for the three countries' recovery plans.[12] Nearly $2 billion had been previously donated following the initial pledge, but the receiving countries were so strapped for resources that they were not equipped to absorb the large sums of money that began pouring in. It opened the door to corruption.

Although the US response was delayed until the middle of the Ebola outbreak, the United States ended up donating the most of any government. In December 2014, Congress approved $5.4 billion for the Ebola response and recovery effort for fiscal year 2015, a larger amount than had been allocated for any previous international infectious disease emergency.[13] The goals of the emergency funding, as outlined by a November 2014 White House fact sheet, were to "fortify domestic public health systems, contain and mitigate the epidemic in West Africa, speed the procurement and testing of vaccines and therapeutics, and strengthen global health security by reducing risks to Americans by enhancing capacity for vulnerable countries to prevent disease outbreaks, detect them early, and swiftly respond before they become epidemics that threaten our national security."[14] An estimated $3.7 billion was designated for international response efforts, $1.1 billion for domestic response, and $515 million for research and development.[15] Before the passage of the emergency Ebola appropriations, US agencies such as USAID, the CDC, and the Department of Defense spent more than $770 million of their already existing budgets. Some of the emergency funding went to reimburse payments made during the initial response.

By September 2016, the US government had paid out a little more than half of the approved money, a portion of which was doled out to external groups through contract grants. In our survey, we asked if US government Ebola stimulus money was appropriately allocated to three different categories: Ebola research, Ebola relief, and future outbreak response. Most respondents believed that funds had been appropriately allocated for research and relief, but 23 percent indicated that not enough money was allocated to groups that were working on projects to prevent future outbreaks, such as strengthening of resources and skills, disease prevention, and drug development.

Several researchers who participated in the survey expressed concern about the terms of the congressionally approved funding. In response to a question asking respondents to describe the biggest obstacle they encountered, one wrote, "Once the outbreak was contained, it was very difficult—contractually—to change to more capacity building and infrastructure for supporting the restart of government-led health care. This was worsened by (1) WHO weakness in coordinating partners and (2) USAID lack of interest in Liberia to learn from the experience and change up its five-year development plan to support new actors and actions." Another stated that the biggest obstacle was "flexibility in funding by USG [US government] to provide most appropriate response to constantly shifting situation on the ground. The USG defined the response strategy, undermining the autonomy of host governments and implementing partners such that the most appropriate resource allocation was routinely stifled. Ebola emergency funding could not be used to build local capacity or invest in efforts to prevent future outbreaks due to an antiquated approach to disbursement of US disaster aid."

It is difficult to determine if these sentiments are warranted, since it is nearly impossible to track the details of how the money was used. Although Congress required agencies to periodically provide reports detailing spending and the progress of the activities funded, the reports have yet to be made public. Little other public documentation is available that would provide clear, comprehensive, and detailed accounting of the funds and the activities funded.

Financial accountability has proven to be among the most challenging aspects of a public health emergency response. This problem extends beyond any one epidemic and any one country. For example, an outpouring of international support arrived in the aftermath of the 2010 earthquake in Haiti and the subsequent cholera outbreak, but minimal tracking of an estimated $6 billion has made it difficult to identify why reconstruction was so slow, and why the public health infrastructure was never rebuilt.[16] There is no public international database that tracks the trail of all money distributed after disasters.

An audit report of the management of Ebola funds by Sierra Leone's Ministry of Health and Sanitation from May through October 2014 tracked more than 84 billion Leones (about $10 million) that was awarded to companies providing goods and services for Ebola response efforts. The money was used for wages for healthcare workers, PPE, and medical supplies. The report found that more than $3 million in Sierra Leone government funds was sent to various hospitals across the country for hazard payments to healthcare workers, but there was no record of the money ever being paid out. Other irregularities included large sums withdrawn from the country's Health Emergency Response Account with no supporting documentation.[17]

The report also found that the biggest contracts awarded by the government did not meet Sierra Leone's regulations for procurement procedures and documentation, such as invitations to bid, acceptance letters, specifications for services and goods, and receipts. Many companies were selected via no-bid, or sole-source, contracts, meaning there was no option for other companies to submit a bid. None of the required documentation, for example, was submitted for Ramesco General Supplies, which was awarded the largest contract, more than $7 million for vehicles and medical supplies. According to the report, Ramesco also did not provide any documentation showing that the supplies it was providing met proper medical standards.

These findings are likely indicative of systemic issues in the country's financial system, yet the effects were amplified during the Ebola outbreak. The lack of accountability almost certainly increased the chances of fraud, waste, and substandard goods and services. The audit report only looked into internal funds—those donated by institutions and individuals mostly within Sierra Leone or derived from tax revenues. It did not address funds that came in through agencies such as the UN and international NGOs because, according to the report, the audit service has "yet to receive information on the quantum of monies received and how they have been expended for the purpose of supporting the response to eradicate EVD [Ebola virus disease]."[18]

"During the Ebola outbreak none of the funds that came in, apart from those that went directly to the government, were subject to the scrutiny of the audit office," said Lara Taylor-Pearce, Sierra Leone's auditor general, in an interview with the Africa Research Institute. "This is despite us writing letters to donors asking to see how the money was being spent."[19] Sierra Leone's audit service sent an official letter to the World Bank in May 2015 requesting details about how its funds and those of other agencies were being spent. The World Bank supervised the distribution of $95 million to UN agencies in Sierra Leone, including the World Food Programme and Unicef, for Ebola eradication efforts. Sierra Leone's health minister, Dr. Abubakarr Fofanah, suggested that external organizations should account for the funds they distributed. "We do not know how these agencies expended the money," Fofanah said in an interview.[20] He also claimed that the World Bank had said it would compensate local health workers, but that the responsibility ended up falling back onto local governments.

Taylor-Pearce said that it is unlikely anyone will ever be held responsible for the lack of documentation and misappropriations. "For serious breaches of financial management procedure I would support forcing the individuals responsible to pay back the money. It would send a strong message," she said. "But currently this does not happen. People continue to get away with transgressions."

There are many reasons why it is difficult to monitor and track global health spending, especially during emergencies, regardless of the donor or whether contributions are private or public. There is no central location or coordinated system where money is gathered, distributed, monitored, and tracked. Further, as more agencies and institutions become involved, the level of confusion rises. There is also no mandate for financial accountability beyond the individual agencies receiving funds, and, thus, no directive to follow the money. International aid may be more easily trackable, since it usually comes in the form of large sums provided to large agencies. But the funds are then divvied up to smaller agencies, contractors, and groups, which are required to track the projects and activities for which they are funded but are not required to publicly disclose how the money was spent. Since no systematic mechanism exists to track investments, and because many are one-time funding allocations, there is no guarantee that long-term efforts to safeguard against future outbreaks are being sustainably funded.

After the Ebola epidemic, some global agencies heeded public demand for greater transparency of their accounts. In 2017, the WHO first published an annually updated online budget and funding portal to offer the public a way to see how much money was obtained and spent, where the money came from, and where the money went.[21] As part of the WHO's compliance with the International Aid Transparency Initiative, the portal identifies which country and program category the money is being allocated to, as well as the type of contributor. The portal, which was established in 2015, covers all of the agency's program funding, not just that for outbreak response. It does not offer detailed tracking of the specifics of implementation. An expanded database that listed the amount and purpose of all funds entering a country would hold governments accountable for the use of those resources.

Accountability is needed for more than donations of financial aid and supplies. There is also a lack of transparency concerning awards and grants that are given for research and development during an epidemic. The Federal Funding Accountability and Transparency Act of 2006 mandated establishment of a searchable website (USAspending.gov) to provide data on federal awards and required that federal agencies report the name of entities receiving awards, the amount received, the recipient's location, and other information. But the database does not reveal the decisions that determine who receives funds, how much they receive, and how the money is spent.

Just as funds intended for aid in Ebola-stricken countries cannot be accounted for, money allocated to fund federal research is difficult to track. In the survey we conducted, many researchers and project managers applying for grants to fund their work found the lack of transparency frustrating. Only 33 percent of participants in our study said they applied for US government funding for their work on Ebola. Of those who did apply, 82 percent said they were able to secure funds. More than half of those who applied (58%) felt they had adequate opportunities to access funding, but most of them also indicated that they had encountered various barriers or challenges when trying to access funds. One of these was the limited scope designated for many of the grants. During a public health emergency, funding is typically directed to immediate needs—such as clinical response or technological support—rather than research with longer-term goals. Our survey asked how easy it was to receive US government funding in 12 different areas: clinical care, diagnostics, vaccines/therapies, biomedical research, clinical research, data analysis, outreach, capacity building, infrastructure, communication, academic research, and industry research. Respondents reported that it was easiest to access funding for projects involving clinical care, and more difficult for those involving infrastructure and academic and industry-related research.

In addition to the problem of limited scope, 40 percent of respondents identified "no open call for applications" as a factor hindering their access to funds; many large grants were earmarked as no-bid contracts for for-profit companies and organizations. In almost every public health emergency, and in times of war, the federal government awards so-called no-bid contracts to companies and researchers to accelerate the process. These contracts are given out when the situation is deemed urgent: disaster is about to strike, or people are dying. No-bid contracts cut the time it takes to review applications from multiple bidders. But some of our survey respondents criticized the no-bid system, saying that it made it significantly harder to locate funding opportunities.

During the Ebola outbreak, about 10 percent of US government contracts were no-bid, a lower percentage than during the Iraq and Afghanistan wars.[22] But how were the companies that were awarded these contracts chosen? In principle, the awarding agency chooses a particular company because it is the only one that has a product that will meet the required needs, or is the only one capable of doing the work during an emergency. Pacific Architects & Engineers Inc. received 16 percent of a $2 billion one-year government contract for Ebola response—likely the largest award in the United States for infrastructure and logistics services.[23] Several pharmaceutical companies also received large contracts. Phoenix Air, a Georgia-based aircraft company, received a little less than 3 percent of the funds. Phoenix Air was the first to conduct flight evacuations during the Ebola outbreak. It supplied the aircraft that carried Kent Brantly and Nancy Writebol from Monrovia, Liberia, to Atlanta for treatment. After those missions, the company was given an emergency sole-source six-month contract by the US Department of State to continue Ebola-related flights from Africa to both Europe and the United States.[24] Phoenix completed at least 40 Ebola-related flights during the outbreak. Although other organizations are developing the capability for high-risk medical evacuations, Phoenix Air claims that it currently has the only aircraft in the world that can successfully complete these types of life-saving missions.

US government agencies are required to provide something called a Justification and Approval document (J&A), which lists reasons for bypassing the normally required competitive bids process, for each no-bid contract awarded.[25] We sought out J&As for all no-bid contracts granted between fiscal year 2014 and 2016. Only one of the 221 no-bid contracts listed as awarded during this period had made its J&A public through the federal online database. Federal Business Opportunities, which provides a database of all types of federal contracts and supporting documents, was unable to provide additional J&As for any of the other sole-source contracts. "It is up to the agency to post that information, it is not a requirement," a service desk agent wrote in a response to our request. Requests to individual agencies did not yield any additional documentation.

There are sometimes good reasons for sole-source contracts. For example, some of them are extensions of contracts that were previously awarded through competitive bids. But although the names of most sole-source contract awardees and the amount awarded are publicly listed, the lack of information about exactly how the money was used makes it impossible to track how the money has been spent, unless disclosed by the awarding agency. Some contracts were made available through the regular competitive process, but the window to apply was so short that it was impossible to solicit fair and open competition.

In the competitive world of scientific research, it is difficult to defend some of the reasons for the no-bid contract; it is rare that one group can provide or undertake a project in a way that no other group can. Unlike the no-bid contract, an open bidding process offers groups the chance to compete, is more transparent, and generally requires more scrutiny and oversight, which can help reduce waste and deter abuse.

Dr. Joseph Fair, senior adviser with the nonprofit research organization MRIGlobal, former special adviser to Sierra Leone's health ministry, and cofounder of Metabiota, says that while no-bid contracts work well when awardees have the capacity to follow through on their purpose, the process can offer an unfair advantage to larger groups, which often receive preference. "We can't innovate in government when we have corporations that have built themselves around the government," said Fair.[26]

High-caliber smaller groups outside of those that typically receive no-bid contracts do exist, he said, and so, too, does the money. But the cumbersome administrative process of applying for grants, and the time it may take to actually receive funds, ends up sidelining smaller groups—ironically, those that need the money the most. "It's not about finding the right people, it is about the time it takes from applying to getting the money, which is a long process," said Fair.

The process of no-bids may cut competition, but it also cuts the time needed for action. Still, Fair believes that no-bid contracts should at the very least be scrutinized with more oversight, and that the bidding process should be modified so as to be fairer and more streamlined.

Before the Ebola epidemic there was no shortage of opportunities to invest in preparedness for epidemics. The CDC, for example, had put in place its Global Disease Detection Program in 2004. Part of the program, which was aimed at detecting and responding to naturally occurring and bioterrorism threats, was to build 18 regional health centers in known disease hotspots around the world to monitor emerging threats. Dr. Scott Dowell, then director of the program, who had supervised the building of seven centers—some of which were only partially functional—appealed to Congress numerous times for funding to build centers in additional areas, including West and Central Africa. According to Dowell, lack of funding is among the primary reasons that, to this day, a center has never been built in West or Central Africa.[27]

The cycle of fighting for adequate funding, access to funds, and calls for transparency in tracking funds continued during the next major outbreak in 2015, when the Zika virus spread from South America to Latin America and the United States. It took nearly a year for the US government to respond, just as had happened with the Ebola epidemic.

In September 2016, nearly 160 infectious disease scientists from all over the world, along with students, faculty, funders, journalists, and the interested public, packed into an auditorium at Boston University for a symposium on the scientific challenge of emerging infectious diseases.[28] It was more than a year after Ebola had receded from the headlines, and Zika, the next public health threat, had taken its place in the spotlight. The challenge, a panel at the symposium explained, was getting the public to care about outbreaks. Pathogens that we know about, and those we don't know about, are spreading faster than the world can contain them. At one point during the symposium, Anthony Fauci, director of the National Institute of Allergy and Infectious Diseases, stood at the podium and flashed a huge map of the world on the screen, with red and blue crisscrossed lines indicating dozens of emerging and reemerging diseases: Ebola. Yellow fever. Zika. West Nile. MERS. Fauci had shown this exact slide to the appropriations committee of the US House of Representatives in March of the same year. "I made this slide completely impossible to read," Fauci said, "because I want to overwhelm them so they will give me more money. For the last few years they haven't, but that's another story."

It's not another story. Indeed, it is the same story recycled during each major outbreak in need of a transnational response. Without proper federal funding reserves in place to predict, prevent, and act on epidemics, researchers and responders cannot access funding until lawmakers steer money to the effort. Funds are not likely to be available until months after the epidemic emerges. In our survey, 12 percent of respondents identified the time it took to receive approval and process paperwork as an obstacle to accessing funds.

The job in West Africa was not finished, but as the public focus shifted to Zika, so too did requests for federal funds. President Barack Obama asked Congress to approve $1.8 billion in new money to combat Zika. Congressional Republicans argued that $2.7 billion of the $5 billion that had been earmarked for Ebola should be reallocated to Zika.[29] The Ebola epidemic was not yet over, and money was still needed for research aimed at prevention and treatment and for efforts to rebuild health systems, but proponents argued that reallocating the money was the fastest way to respond to the new health threat. The Senate had approved $1.1 billion in new funds for Zika in May 2016, and the House soon afterward approved $622 million, far below the amount requested. This led to continued disagreements, which only further delayed funding. "It is now September 18, [2016]," Fauci told the crowd at the Boston University symposium, "and we still don't have [the money]." A little more than a week after Fauci's statement, Congress agreed to allocate $1.1 billion for Zika.[30] While waiting for lawmakers to act, President Obama had redirected $510 million of Ebola funds to be used for Zika instead. The money left the Ebola fund and was never replenished.[31]

The US funding strategy implemented during the Ebola and Zika epidemics suggests that an even larger human and financial cost would be incurred if a deadlier threat emerges. Funding delays hampered efforts to understand and stop the outbreak. Delays in funding are known to waste lives and time and to end up costing more. Indeed, in the case of Ebola, and most other large-scale epidemics, funding readiness is a more cost-effective strategy than funding the response. However, the challenge has been to convince government officials to allocate funds in advance for anticipated threats. Funding for global health initiatives steadily increased under the George W. Bush and Obama administrations, but the majority of the money was allocated to fighting high-profile diseases like malaria and AIDS. Only a sliver has been directed to broader prevention-based surveillance and response programs.[32]

In 2015 alone, the three core countries in West Africa that were hardest hit by Ebola lost a combined $1.6 billion due to the outbreak, over 12 percent of their combined GDP.[33] The World Bank estimates that Sierra Leone lost $163 million in GDP in 2014 alone due to Ebola. Guinea and Liberia lost $130 million and $66 million, respectively.[34] The longer it takes to get a coordinated international response, the larger the economic devastation. The severe economic impact to countries and international responders during an epidemic is not unique to Ebola.

Mitigating the economic impact of an epidemic requires that all countries invest in pandemic preparedness. Money is usually not discussed until after the emergence of an epidemic, and as we know, by then it's too late. Funding delays devastate efforts for a more streamlined response and make it especially difficult for those who need money for a specific purpose—whether for research or rapid response.

"We should have responses ready to go within hours or days instead of months and years, as we had [with Ebola]," said Jeremy Farrar, director of the research charity Wellcome Trust. "It's got to be properly funded and organized and ready to go tomorrow, not in a year's time."

The world cannot rely on the emergency funding measures that are currently in place. After the Ebola epidemic, the WHO created a contingency fund to provide rapid funding in emergencies. By 2017, it had attained only about 30 percent of its $100 million goal, and most of that money was already committed to existing emergencies.[35]

The World Bank launched a first-of-its-kind "pandemic bond" in 2017, intended to help raise funds in 77 of the poorest countries in the world in the event of an infectious disease outbreak. The bonds, part of a global financing mechanism known as the Pandemic Emergency Financing Facility (PEF), are intended to provide a way to mobilize funds quickly. The system is essentially an insurance policy, paid for by donors who purchase the bonds, that would provide up to $500 million in coverage for a five-year period against outbreaks identified as likely to cause major epidemics.[36]

Dr. Michael Osterholm, director of the University of Minnesota's Center for Infectious Disease Research and Policy, has called the plan a "bankrupt concept" because it emphasizes funding for response over readiness. Many of the countries that qualify for the pandemic bond lack the infrastructure and the level of preparedness needed to effectively prevent epidemics. "It may help in regional conflicts," said Osterholm. "But when you have a pandemic, it's going to be useless." Part of the problem is our mindset. "If we wait until a crisis happens, we're way too late already."[37]

Financial challenges will continue unless the world realizes that natural disease outbreaks are just as dangerous as the intentional release of pathogens in biological warfare. At present, demand for research and development increases—along with financial resources—when a particular pathogen is identified as having the potential to be used as an agent of bioterrorism. After the terrorist attacks on the United States on September 11, 2001, several governments around the world invested in research on Ebola because of concerns that the virus could be used as a weapon.[38] Interest in the potential use of Ebola as a biological weapon increased again after the 2014–2016 Ebola outbreak.[39]

Significant funding is available for defense against bioterrorist attacks. The United States set aside emergency funding under the Project BioShield Act of 2004 to defend the country against potential attacks from chemical, biological, radiological, and nuclear (CBRN) agents. As part of Project BioShield—a 10-year program—$5 billion was put into a "special reserve fund" to purchase critical medical supplies, such as vaccines, drugs, and diagnostics. These funds are doled out when either the secretary of health and human services declares a public health emergency that has a significant potential to affect national security, or the secretary of homeland security or the secretary of defense determine that there is a heightened risk for a CBRN attack. The purpose of Project BioShield is to "accelerate the research, development, purchase, and availability of effective medical countermeasures against CBRN agents."[40]

The challenge of implementing the equivalent of Project BioShield for natural infectious disease outbreaks is that there are so many of them. But this is the kind of approach that is needed. Many scientific leaders in public health emergency response feel strongly that it is time to establish a permanent global health security fund dedicated to supporting preparedness actions such as building public health infrastructure, undertaking surveillance projects, and developing diagnostics and therapeutics for global health threats.

All of the issues discussed in this chapter—lack of financial accountability, obscure methods of allocation and tracking, inadequate funding for capacity building efforts between epidemics—present fundamental challenges for outbreak response. Lobbying for money during an emergency is a time-consuming and costly exercise that should be replaced with continuous support for infectious disease prevention, surveillance, detection, and response. A public health safety net is essential to prepare for another infectious disease outbreak, as is a global system to ensure accountability for where and how the money is spent. While funding is vital, no amount of money would improve outbreak response without systemic changes, revised regulations, and enhanced tracking. The global community's reluctance to improve funding practices shows that we are continuing to be reactive, not proactive, to global health threats—a habit that has repeatedly proven to be a deadly mistake.

8

EBOLA'S FALLOUT

Epidemics ordinarily end with a whimper, not a bang. Susceptible individuals flee, die, or recover, and incidence of the disease gradually declines. It is a flat and ambiguous yet inevitable sequence for a last act.

—Charles E. Rosenberg, "What is an Epidemic?"

BY SPRING OF 2015, the Ebola epidemic began to die down in West Africa. First Guinea, then Sierra Leone, then Liberia. In each country, the WHO declared the epidemic over after 42 days went by with no new cases.[1] Liberia was declared free of Ebola in May 2015.

But containment does not mean eradication. The Ebola virus still circulates within the three hardest-hit countries, and new cases appeared in Liberia twice after the end of the epidemic was declared. As in the case of most epidemics, medical workers must vigilantly watch out for residual cases. Many response leaders on the ground, including Dr. Matshidiso Moeti, WHO regional director for Africa, cautioned that the resources that poured in and the progress made in building infrastructure, response systems, and skills had to be sustained to keep Ebola contained.[2] By August 2015 public interest in Ebola seemed to be waning, and international responders were suffering from epidemic fatigue. The world marked the end of the epidemic, but the outbreak, and all it brought with it, was far from over.

Dr. Joanne Liu, international president of MSF, was among those who criticized leaders, health agencies, organizations, and academics for prematurely turning their attention to retrospective analyses of the outbreak and generating documents of "lessons learned." "I have witnessed how a lack of political will undermined the response in the early days of the epidemic," Liu wrote in *Nature* magazine. "Now, fatigue and a waning focus are threatening the final push to the end."[3]

By the end of July 2015, most European responders had completed their missions, including the French military and Portuguese government teams, which had returned from their clinical and laboratory building efforts in Guinea.[4] A skeleton unit of NGOs and foreign aid workers remained in West Africa as the region experienced the fallout from Ebola.

The epidemic decimated the three countries' already depleted health systems. Maternal and child health saw the largest collateral impact. Death rates of mothers and infants were already very high—in Guinea, for example, the mortality rate for children under the age of five is 1 in 10, and the lifetime risk of maternal death is 1 in 30.[5] During the Ebola outbreak, governmental and nongovernmental agencies in all three countries launched health education campaigns encouraging mothers to seek conventional medical care. Many women who heeded the call were turned away due to lack of resources and overcapacity. There wasn't much that could be done for women and infants who contracted Ebola. Pregnant women with Ebola were at an increased risk of dying during childbirth from hemorrhage. Their babies, if not stillborn, would die shortly after birth as a result of a high viral load or of contracting the virus through breast milk.[6] Exposure to blood and other fluids during childbirth also put caregivers and even other expectant mothers in the ward at increased risk.

The World Bank estimates that the loss of healthcare workers alone led to around four thousand additional maternal deaths in health centers across Guinea, Liberia, and Sierra Leone.[7] This is a conservative estimate, since many pregnant women to this day refuse to visit hospitals and health centers because they are scared of contracting Ebola.

Childbirth was not the only area of increased risk that West Africans faced during the fallout from Ebola. The focus on diagnosing and treating Ebola in resource-strapped medical centers led to a drop-off in care for a variety of common diseases, including malaria, tuberculosis, HIV / AIDS, cancer, diabetes, and heart and lung diseases.[8] It is estimated that 10,900 more malaria deaths occurred in the three countries in 2014 than if there had been no outbreak.[9]

Ebola has taught us that the anthropology of a region should always be taken into account during outbreak response. The region's history of colonialism and political instability led to suspicion, and in some cases aggression, against foreign and governmental responders. Our study findings suggest that to residents in the affected regions, the reaction to Ebola was determined largely by how the disease affected their community structure and way of life. Some of the respondents to our survey suggested that the key to restoring trust was to make sure that messages and interventions were led by local groups. "Local community members should be involved in response activities and planning," wrote one survey respondent. "This will give them more confidence in taking responsibility for the protection and health security for their communities." The respondent added that this increased confidence would benefit both local communities and international responders.

Regaining community trust can also be fostered by building up regional health systems and making them sustainable. Doing so requires committing aid for the long haul.[10] A temporary influx of clinical workers and medical supplies during the Ebola outbreak was not enough to foster trust or to create comprehensive treatment plans for chronic ailments. Many local people questioned the sincerity of response teams that focused only on Ebola cases while ignoring other health issues. Continued lack of trust jeopardizes long-term containment practices. In Guinea, for example, hostility to outbreak responders remained strong for the duration of the epidemic and beyond. In May 2015, more than a year after outbreak responders first arrived in the country, the Red Cross had to withdraw parts of its team after two cars and an employee's home were attacked, and a warehouse containing equipment for performing safe burials was burnt down.[11]

As the epidemic wound down, international agencies reduced the level of supplies and personnel being directed to the region. Clinical responders were either sent home or moved on to other regions with public health concerns. A majority of local government health workers at the ETU in Foya, Liberia, were laid off after the number of Ebola cases tapered off. Darlington Jallah, the head nurse in Foya who knew every Ebola patient's name, said the workers did not receive their final month's pay and were treated as if they had never existed. "Their names have been deleted from the government payroll," he said.[12]

The health workers, said Jallah, could have been kept on to care for the community's broader needs in the aftermath of the deadly outbreak. He described the staff as "downhearted," having risked their lives working for the government at the MSF-run unit only to have their jobs taken away and their contributions unappreciated. Some workers lost confidence in the government's ability to follow through on promises that it would care for its health workers in the aftermath of the traumatic situation. Jallah himself has since left Foya. He went back to school, earning a master's degree in public health, and has been looking for a new job.

Many initiatives that were undertaken during the response had the potential to lead to long-term, systemic changes. As supplies and personnel arrived in the region, effective infection and prevention control measures were implemented that not only reduced transmission of Ebola but helped local people develop good hygiene practices, including handwashing, safe burials, and proper handling of infected patients. Sustaining these habits, however, will require additional funding.

An enormous amount of monetary aid was wasted because restrictions prevented funds from going to improvements not deemed immediately essential for the Ebola response. Thus, much of the funding went for the construction of temporary structures, the recruitment and training of temporary staff, and the delivery of Ebola-specific services. It was almost impossible to secure financing to strengthen primary care services, such as support for maternal and mental health, as a component of an Ebola response grant.

All medical services, even those for basic illnesses, deteriorated after the epidemic because so many doctors, nurses, and midwives had died. During the outbreak, healthcare workers were 21 to 32 times more likely to be infected than the general population. Before the epidemic, Sierra Leone, Guinea, and Liberia had ratios of about 1–2 doctors for every 100,000 people, among the lowest in the world (the recommended level is 250 to 100,000). Sierra Leone lost over 220 health workers by January 2015, and the WHO recorded 157 deaths in Liberia and 109 in Guinea between January 2014 and March 2015. In Liberia, 8 percent of all healthcare workers died of Ebola.[13]

Thousands of people died from Ebola during the epidemic, but thousands also survived. The health problems of survivors have been described as an "emergency within an emergency" in West Africa.[14] Continuation of care was not planned as part of outbreak response and so these problems have not been well documented, but it is clear that many survivors lack adequate clinical or supportive care. We do know that survivors often suffer from what has been called post-Ebola syndrome, a collection of symptoms that include musculoskeletal pain, headaches, and vision loss.[15] One project has been successful in helping some survivors. A group of NGOs in Sierra Leone opened a survivor eye clinic in the Port Loko district in 2015. The Ministry of Health and Sanitation took over the program and expanded it into a national eye program. This program may well have prevented an epidemic of blindness among survivors.[16]

Some survivors felt deceived when promises of help were not met, and the designated funds were unaccounted for. In December 2017, two Ebola survivors filed a lawsuit against the government of Sierra Leone, alleging that the lack of government accountability and mismanaged funds denied citizens the "right to life and right to health."[17] The case, filed with a regional court in Nigeria, is the first international court case to focus on accountability for missing and misappropriated funds associated with the Ebola epidemic. Ibrahim Tommy, executive director of the Freetown-based Centre for Accountability and Rule of Law, which supported the plaintiffs, and Yusuf Kabbah, head of Sierra Leone's Ebola Survivors Association, said in a news conference that government health officials had promised Ebola survivors compensation for livelihood and healthcare, and access to psychosocial support—none of which were ever received.

Horrific as the epidemic was, it directly affected only a small proportion of the populace: it is estimated that about 0.25 percent of the population was infected with Ebola in Sierra Leone and Liberia, and less than 0.05 percent in Guinea.[18] Despite these relatively small numbers, the costs of health services during the epidemic were substantial, and indirect costs on other sectors of the economy proved just as significant. Ebola devastated the already struggling economies of the three core countries. Economic recovery in Sierra Leone has been slower than in Guinea and Liberia, since it was the hardest hit.[19] Food production and distribution declined because of restrictions on movement and closed markets, and prolonged school closings set back the educational progress of many children.

Fear was another consequence of the epidemic. Survivors include not only those who recovered from infection with the virus, but those who witnessed the societal breakdown and trauma that accompanied the outbreak. There was an upsurge in the incidence of mental health disorders, including depression, anxiety, post-traumatic stress, and substance abuse. Many people never received the help they needed. Groups in need of mental health treatment included caregivers and those who lost friends and family members. We may never know the actual number affected by the epidemic because of the sense of shame people felt about seeking help. Members of families that were torn apart by the epidemic still harbor anger and suspicion toward Ebola responders and even toward neighbors who reported suspected cases. In Sierra Leone, Fambul Tok, the organization that had initiated truth and reconciliation forums among communities affected by the civil war, began to turn its attention to regions ravaged by the epidemic.

Many West Africans resented foreign responders and the decisions they made. Emergency response often demands quick action, and during the Ebola epidemic, decisions at every level from the global to the regional and local were most often made single-handedly by foreign responders. Some Africans as well as foreign responders saw the power dynamics at play during the outbreak as a continued manifestation of colonialism. One of our survey respondents wrote: "Racism and injustice should not be common features of an outbreak response occurring on a continent of black humans. Politics should not play a role in resource allocation. Hoarding precious resources in the US 'in case' of a domestic epidemic by hundreds of hospitals which will likely never see an Ebola patient created a false shortage of these resources in West Africa."

Those who have recovered from Ebola face the most serious consequences, often as a result of being stigmatized by their communities. Some have been turned away from their homes or workplaces and are shunned by their friends. MSF opened short-term survivors' clinics in all three countries immediately after the epidemic, but ongoing services are needed to address continued social and mental health issues. In 2015, the US National Institutes of Health, in conjunction with the Liberian government, established a large-scale, five-year project to assess long-term health effects in Ebola survivors in Liberia.[20]

The work of eliminating stigma and reintegrating survivors is critical to the future of Ebola response in part because survivors could be a valuable resource in any future epidemics. They could help to mobilize communities and also provide direct care of Ebola patients. Preliminary evidence suggests that some survivors have at least short-term immunity from reinfection and may be able to care for Ebola patients without the protection of PPE.[21] They could be especially useful in caring for children and young adults, who are more likely to suffer from long-term psychological impacts than are adults.

Building and rebuilding health systems will take a long time, probably more than twice as long as the epidemic itself lasted. The governments of Sierra Leone, Liberia, and Guinea all introduced strategic plans in 2016 to fix their countries' health systems, including strengthening scientific research programs. The government of Guinea put together a $2 billion post-Ebola recovery plan, with 63 percent of the funding allocated to basic health needs, nutrition, and hygiene.[22]

The effects of the epidemic were felt far beyond the countries where the outbreak occurred. In the United States, the CDC undertook the largest mobilization in its history. Four thousand CDC employees were involved in the response, including 1,400 people who spent a total of 75,000 days on the ground in West Africa, performing contact tracing, infection control, health communication, incident management, and laboratory work.[23]

The wider US population was also affected, as valid concerns about the possibility of an outbreak on US soil quickly exploded into unjustified panic. Any news about Ebola cases in the United States would probably have generated public anxiety, but that anxiety was heightened by the inconsistent messages coming from experts.[24] Management of the Ebola "crisis" in the United States taught us that it is more important for experts to communicate early and transparently, even if it means conveying uncertainty, than to offer firm predictions during an unpredictable situation.

The Ebola epidemic taught us—yet again—that an outbreak anywhere in the world can cross oceans, desert, and forest to become an outbreak everywhere. The pathogens that cause these outbreaks do not have to be viruses—they can be bacteria, parasites, or fungi. They can be existing pathogens or new ones, like SARS and MERS. They can spread naturally or, perhaps, be deployed as biological weapons. They can be so pervasive that they destabilize political, economic, and social infrastructure for decades, like HIV. None of them are subject to human-drawn borders, thus forcing us to consider responses that go beyond placing restrictions on travel or quarantining communities.

Ebola taught us that what limits our response to outbreaks is mostly our collective mindset and our own management and behavioral failures. "The limitations are human, they're not technical anymore," said Nathan Yozwiak, associate director of viral genomics at the Sabeti Lab. Early in the Ebola epidemic, not much was known about the particular strain of Ebola that was spreading through communities in West Africa. Epidemiological research—that is, understanding the origin of a disease and its spread—is typically done retrospectively. Epidemiologists wait until the dust settles to learn about the situation. In Kenema, the Lassa Fever Research Program had existed long before the Ebola outbreak started and was well integrated with clinical care efforts to treat patients. Dr. Khan's goal was to follow the same model so as to understand the Ebola virus in real time, even as the number of patients grew. But as more and more outsiders entered Kenema and tried to control the operations, this led to chaos, confusion, and mistrust in a high-risk environment, followed by restrictions on research and hoarding of data. The toxic environment made systematic research during the outbreak challenging, when instead it should have been regarded as a critical component of outbreak response.

The Belgian microbiologist Dr. Peter Piot said in an interview that collaboration during clinical trials of new vaccines could also have led to more rapid discovery. The London School of Hygiene and Tropical Medicine, where Piot is the director and professor of global health, was one of a handful of institutions and global agencies that was working with governments to conduct trials of candidate Ebola vaccines in 2015. The London institution supported the WHO vaccine trial in Guinea and led trials in Sierra Leone with the pharmaceutical manufacturer Johnson and Johnson. Trials in Liberia were overseen by the National Institute of Allergy and Infectious Diseases, part of the National Institutes of Health. "In an ideal world we should have had one trial with the two or three vaccines that had some reasonable consideration for further development and we would've had the answer sooner," Piot said. "I don't think it had to do with corruption or people wanting to be first," he added. "It's about disorganization and lack of preparedness. So the next time we have to make sure we already have some agreements beforehand."

Clinical and technological advances such as drugs and vaccines offer promise for future outbreaks but were unavailable in large enough quantities or with enough certainty to have made a substantial difference during the epidemic.[25] Rapid diagnostic tests, for example, which can deliver results in a few minutes and require only a small amount of blood, were still being studied. Systematic coordination of data collection and research would have advanced understanding of the virus and the ability to detect and combat it. Results from the studies that did go on during the outbreak attest to the importance of continuing to do research even in a chaotic and unpredictable environment. Continued sequencing of the virus genome and immediate dissemination of the data helped to determine the path of the outbreak and identify the mutations that made the virus more infectious. Trials of the drug ZMapp that were conducted in West Africa during the second half of the epidemic helped establish its therapeutic benefit and probably saved some lives.[26]

Continued research efforts in the post-outbreak period will help to ensure that effective therapies are available the next time they are needed. The process can be sped up by streamlining approvals from multiple governments to research, test, and approve therapeutics for future use. Fast work on new vaccines made it possible to use an experimental vaccine during a small flare-up of Ebola in the Democratic Republic of the Congo in 2017, and a subsequent outbreak there in 2018.

It would be imprudent, however, to rely solely on the outcome of clinical research, such as vaccines and drugs, to curb an epidemic without an effective warning system and centralized response system in place. Hundreds had already contracted Ebola and died in the five months between the first formal notification of the virus and the declaration of an international public health emergency, which launched the full-scale international response. Help, when it arrived, was bountiful, but the entire episode proved that timing is everything. It took many months from the time the United States announced that it would deploy three thousand troops to Liberia for the ETUs to be built. By then, the epidemic had waned, and 9 of the 11 centers never saw an Ebola patient.[27]

Since many actors are involved in events that warrant large-scale operations, collaboration is essential for a speedy response. The WHO's International Health Regulations, a treaty established in 2005, requires countries to prevent, detect, and respond to outbreaks, but as we discussed in Chapter 4, it had little effect during the Ebola epidemic because of lack of compliance. The Global Health Security Agenda (GHSA), launched in February 2014 by a consortium of 29 nations, international organizations, and other public and private partners, is one example of a large-scale collaborative response intended to improve outbreak response.[28] The idea behind the initiative is for countries to share resources with participating agencies and countries during a disease outbreak, with the goal of identifying threats and responding quickly, putting out small fires before they spread. Unfortunately, this initiative came too late to play a role in the Ebola epidemic.

Action plans among many international bodies—including GHSA—as well as governmental and nongovernmental organizations include capacity building, training healthcare workers, and strengthening public health systems. Rarely included on the agenda of improving outbreak response, however, are the important goals of ensuring accountability and eliminating corruption. Ebola has taught us that we have reached the limit of what classic containment can achieve. As we discuss in the next chapter, our global health security depends on our ability to look beyond the pathogen itself and toward bolstering health systems, increasing cooperation, and establishing a centralized governing structure before the next epidemic.

Just as the epidemic waned in West Africa in 2015, governments and international public health agencies saw the opportunity for improved response with the emergence of a different virus. The mosquito-borne Zika virus was detected in Brazil and led to an outbreak of Zika fever that spread into other parts of South America, Central America, and the Caribbean. In February 2016, the WHO declared the outbreak a public health emergency of international concern, after cases of microcephaly and other neurological disorders were found in infants born to women who were thought to have been infected with the virus. The WHO was breaking precedent by making the declaration before the connection between the virus and the birth defects was confirmed. The CDC also elevated its emergency response to its highest activation level, boosting US government–led surveillance and research. President Obama requested $1.8 billion for Zika response within a week, marking a sharp contrast with the situation during the Ebola epidemic: in that case, Obama did not request emergency funding until three months after a global emergency was declared. It took only about three and a half months after the release of the Zika virus's genomic sequence for researchers to begin work on a vaccine, a milestone in the history of vaccinology.[29] Multiple vaccine candidates developed since 2016 show promise in preventing Zika infection, but momentum to bring one to market has slowed as cases have declined and public attention has turned elsewhere.

Despite the fast response to the Zika outbreak, many of those involved noted that the same features that plagued outbreak response during the Ebola epidemic were still at play. Although the WHO encouraged government, academia, and industry to share data, for example, it came as a request, not a requirement. Drug companies developing Zika vaccines were among those who did not agree to participate in data sharing.[30]

Although the perception is that swift action was taken with Zika—and that may be the case for this particular outbreak—the reality is that Zika is not a new virus to emerge on scientists' radar screens. The virus was first discovered in Uganda's Zika forest in 1947 and has caused outbreaks in Southeast Asia and the Pacific Islands in recent years. Even though scientists had known about it for decades, it was not considered dangerous, and no diagnostic test, treatment, or vaccine had been developed. The response to the outbreak could have been more effective if a research plan had already been put in place to detect the virus's evolution and migration, keep track of cases, and develop therapeutics.

Zika and Ebola highlight the persistent flaw in a global public health system—the focus is primarily on response. Nearly every research paper discussing Ebola before the 2014 outbreak ended with a warning about the likelihood of future epidemics, yet that warning was overlooked. We know now that we have not seen the last of the Ebola virus. Africans will continue to contract and die from Ebola during flare-ups. In September 2014, a team of ecologists and epidemiologists at the University of Oxford created a map of the places where Ebola was most likely to emerge in the future. The high-risk area consisted of a thick belt across central Africa, including Tanzania, Mozambique, and the island of Madagascar, with a total population of about 22 million people.[31] Local health workers bear the brunt not only of caring for survivors but also containing small outbreaks, all while working to prevent the next large-scale epidemic.

The 2014–2016 Ebola epidemic will not be the last epidemic of its size and scope. New pathogens are constantly being discovered, while old ones reemerge. Some diseases die out as quickly as they began, while others reach epidemic and pandemic proportions. The course of any epidemic varies as a result of characteristics of the human population, the environment, and the pathogen itself. While scientists can use these factors to predict future outbreaks, it is unrealistic to keep track of every emerging pathogen because the benefits are not worth the effort. After all, most outbreaks are limited to a few people or a single remote village. While the global community could do more to surveille human populations, monitor, and detect changes to determine the potential for large-scale outbreaks, it is far too ambitious for scientists to try to anticipate the exact course and magnitude of the next Ebola outbreak or that of any other specific pathogen.[32]

Key features of previous epidemics can help to guide future forecasts. One study that examined more than 200 human infectious diseases and more than 44 million cases over three decades found that both the number and the diversity of outbreaks have significantly increased globally since 1980.[33] With increased urbanization and changes in the environment, many subsequent outbreaks will likely be zoonotic—transmitted from animals to humans.[34] Approximately 60 percent of all human pathogens are zoonotic, and 75 percent of recent emerging infectious diseases that affect humans come from animal reservoirs.[35] It is difficult to predict which, if any, of these pathogens will create an epidemic. However, knowing the likelihood has assisted with global health agencies' and governments' preparedness. In West Africa, leaders agreed to a new collaborative approach after the Ebola epidemic. Their health ministries are addressing human, animal, and environmental health collectively. This is a large feat for countries with already strapped health systems.

The World Health Organization described the 2014–2016 Ebola epidemic in West Africa as an "old disease in a new context."[36] Despite the fact that the Ebola virus is not highly contagious, it was still able to infect and kill thousands. The situation could be much worse in the case of a virus like influenza that is spread by airborne transmission. Even familiar pathogens can become more virulent as new strains evolve and our environment changes.

In the future we will probably see more outbreaks because of population growth, degradation of the natural environment, and climate change. We are getting better at detecting them, but it is how we prepare and react that determines their course. What is being done to protect against an even more virulent epidemic? Based on what we learned during the Ebola epidemic, the answer is: not much. If we respond to the next outbreak in the same way that we responded to the Ebola outbreak in West Africa, we will have failed to learn from our mistakes. The next epidemic demands better.

NAVIGATING THE NEXT EPIDEMIC

Knowing is not enough; we must apply. Willing is not enough; we must do.

—Goethe

THERE IS A GOOD CHANCE that the world will see at least one pandemic during the coming century, and a 20 percent chance of four or more.[1] Standard measures of outbreak control will not be adequate to handle these future health crises. The more contagious the pathogen, the more difficult it will be to contain. And, as we saw with Ebola, the more time the virus has to spread from human to human, the more infectious it can become. To prepare for this threat, we need to shift outbreak response to a mode that favors collaboration instead of competition, and readiness instead of reaction.

In the wake of the Ebola epidemic, many global health agencies and public health groups have already offered concrete recommendations for changes in practice and logistics that could improve outbreak response.[2] Other groups have proposed legal reforms, mainly at the state and local level. These recommendations and models of legislation address key public health measures to enhance future preparedness, surveillance, capacity, and communication, with the ultimate goal of better protecting the public.[3] We support these recommendations, and it is not our intent to replicate or supplant them. We aim, instead, to set forth a number of guiding principles, based on our own experience and research, to reframe how the global community views disease outbreaks and epidemics.

CONFRONTING HARD REALITIES

Improving outbreak response first requires acknowledging the extraordinarily challenging conditions that accompany any disease outbreak. As we have seen throughout this book, outbreaks are a crucible. The environment is volatile, the situation is often unclear and rapidly evolving, and lives are at stake. Every person in the community is perceived as a potential threat, leading to an environment of mistrust, paranoia, and suspicion. Blame and stigma thrive. Economies are devastated, and civil governance breaks down. At the same time, there is political pressure not to acknowledge the crisis until it is impossible to ignore.

The international community's approach to outbreak control currently relies more heavily on response than readiness. Disproportionate dependence on emergency intervention has not only left us more vulnerable to naturally occurring infectious diseases, but has allowed the destructive aspects of outbreak culture to proliferate. In the chaos of rapid response, a multitude of international groups hastily move into a region and do not have or make the time to develop positive working relationships and trust with local communities or with existing public health teams.

A number of perverse incentives can also take hold. Outbreaks can boost scientific research, advance political careers, and generate profit for private companies, sometimes at the expense of civilians on the front lines. The response to an outbreak brings in large amounts of cash, creating opportunities to exploit gaps in accountability and jockey for recognition.

The tension between the stressful environment and the desire to gain funding or renown, combined with the chaos and confusion, can lead all actors to fall prey to a toxic culture. Sometimes actions that contribute to the detrimental environment are accidental. Other times they are coincidental or cultural. But we must also acknowledge that there are intentional bad actors in every outbreak. These actors try to take advantage of any weakness in the system and can poison interactions between all of the elements. Individuals or groups may exploit fears for political gain. Researchers may attempt to propel their careers and secure prospects for future funding by using illegal or unethical tactics to intentionally block collective progress.

We need to eliminate the barriers that are impeding organization and collaboration. It is possible to improve outbreak response, but it must be a deliberate choice, thoughtfully implemented, governed by core principles, and guided by realism, honesty, transparency, and respect for others.

ALIGNING INCENTIVES

Acknowledging the hard realities of an outbreak situation is the first step toward creating a system that works for all parties. We must recognize that each individual actor has their own set of concerns and needs and then attempt to align incentives such that cooperative behavior is rewarded. "For me, the key going forward is to change the incentives in the system so we benefit the people directly affected by the outbreak," said Belgian microbiologist Dr. Peter Piot.[4]

Paramount to improving preparedness and early response is to report and share information, samples, data, and discoveries in as close to real time as possible. Patients must seek care when they have symptoms, and scientists must make data available to others. To encourage people to behave in these ways, we must consider what causes them not to do so. Infected patients will only seek care if confident that they will be treated well, rather than being forced into quarantine and stigmatized. Researchers can only be convinced to share data if they are recognized for their effort and allowed to pursue their projects, rather than being exploited and left without support. Fear of exploitation, for example, is behind the idea of viral sovereignty, which regards viruses as natural resources belonging to the country in which they were first discovered.

The current system of patents as it relates to outbreaks must also be revised, so that all individuals have incentives to engage in the effort to prevent and treat infectious diseases. Research and data dissemination should be guided by the principle that a pathogen does not belong to any one country, or to any one group or person studying it. New discoveries should be made available to everyone, especially those most affected by the disease.

Aligning incentives should be considered in all aspects of outbreak response. For example, we have seen time and time again that individuals resist quarantine practices that follow a punitive model, where people are trapped in uncomfortable environments and stigmatized. Participation and compliance are far more likely if quarantine and monitoring are made comfortable and are treated as a partnership between public health experts, the community, and the individual.

ASPIRING TO ORGANIZATIONAL JUSTICE

Any system put in place to manage future outbreaks must begin with a principle of organizational justice that extends to every individual and actor in the system. If the global response effort is perceived to be fair, transparent, and just, individuals will remain loyal and trusting. As we have seen in many outbreaks, including the Ebola epidemic, perceptions of injustice breed distrust and misbehavior.

Our study findings suggest that outbreak responders need to feel confident that the global health community is actively working to hold people accountable for willful misconduct or gross negligence. We found that illegal and unethical tactics, and even actions perceived to be as such, during the response were a major obstacle to a coordinated and collaborative response at all levels.

Many of our sources, as we discussed in Chapter 2, blamed the lack of central governance, transparency, and accountability for everything from general discord in response efforts to flagrant systemic corruption. When many actors from various countries and agencies are involved, we need a well-defined and universal ethics to approach outbreak response. An international legal framework that lays out who is responsible for what during outbreaks would be beneficial, as it would provide standards that are clearly and explicitly articulated, written down and agreed upon, and followed by everyone within the global community. All actors involved in outbreak response, regardless of size and scope, should be held accountable for their actions and outcomes.

While responding to outbreaks requires a centralized global effort, it is the people within the affected communities who are most important for detecting the emergence of disease and sounding the alarm. Even if local community members do not know the name of the ailment that is causing a growing number of illnesses or deaths, they will be willing to report their observations if they have formed relationships with healthcare workers. Building an effective response requires that individuals feel safe when reporting changes within their community. This level of trust, awareness, and reporting requires engagement not just during outbreaks, but during interepidemic periods as well.

TRUTH AND RECONCILIATION, FAMBUL TOK

There is currently no centralized system for reporting misconduct, and no agency that handles these occurrences across the spectrum of responders. Disruptive practices are typically not identified until a retrospective analysis of the situation is completed, when it may be too late. We need to form an international system to regulate outbreak response and move the culture in the direction of transparency, collaboration, and accountability.

Although a majority of responders typically embrace a set of self-directed principles out of the desire to help the common good, a legal framework is necessary to ensure regulation and compliance. The system should reward responders who work collaboratively. Any actor—local and national governments, clinicians, scientists, relief agencies—involved in infectious disease outbreaks should be held to universal standards defined by the global community.

To be effective, a reporting system has to encourage individuals to report infractions and protect them when they do so, while at the same time discouraging, or at least not rewarding, false reporting or any other behavior that would subvert the system. The system should follow a model based on mediation and shared solutions rather than punishment, except in the most extreme cases.

One approach is to solicit real-time feedback from responders, as a way to "check the temperature" of the outbreak response system. To ensure communication from a broad number of responders and identify problems early on, systematic surveys should be undertaken when responders leave a host country, after they have had time to reflect on their experience as a whole.

The global community could learn a great deal from the Sierra Leonean tradition of rebuilding communities by candidly confronting issues within a safe setting. The organization Fambul Tok ("Family Talk"), which provided forums to strengthen communities devastated by the decade-long civil war, should be considered as a model during the sensitive post-outbreak period.[5] Honest personal accounts could help everyone affected by an outbreak to come to terms with their experience. Coming together to forgive and heal promotes a more compassionate culture.

READINESS

Readiness refers to a state of perpetual preparedness, a concept that many experts believe the global community has yet to grasp. A lack of readiness is the reason that the chaotic international response during the Ebola outbreak caused efforts to initially slow down rather than speed up. While Ebola was successfully prevented from becoming a pandemic, it cost more lives and dollars than necessary. By the time an outbreak reaches the epidemic phase, it is too late to create an effective response system.

Plans should be laid between, not during, epidemics. The more time that passes between outbreaks, the less likely we are to implement the lessons learned. But planning is not enough. A smarter approach to readiness involves commitment to continual health promotion and long-term community engagement to bolster health systems during "quiet" periods—that is, periods when public attention is at a low ebb. This strategy is part of what Boston University public health professors Sandro Galea and George Annas describe as a "common-sense public health agenda," where global leaders in public health shift toward promoting population health and away from reacting to public health emergencies.[6] Building on existing public health efforts during an outbreak will result in a more streamlined and adaptable effort.

In most cases, there is only a difference of scale between the components needed to respond to an isolated case and those needed to contain an outbreak. Caring for one case of Ebola is much the same as caring for multiple cases of the disease, with the exception of added supplies and personnel. Thus, methods established and practiced during the interepidemic period will already be familiar to responders during an emergency.

Seed Global Health, a US-based nonprofit organization that trains local health workers in countries with underdeveloped health systems, follows what its CEO and cofounder Dr. Vanessa Kerry calls a "time-intensive approach to solve problems." Kerry fears that our current culture, which prioritizes response and quick fixes, does not consider what it takes to address major issues, especially since a majority of these problems require prolonged time to tackle.[7] Her organization's approach is to rely on funding from both public and private sources over periods of time longer than the typical cycle of a health disaster.

Organizations like Seed Global Health build capacity and trust during quiet times. But the global community cannot rely on individual organizations or even a few philanthropic funders, each conducting its own mission. One health worker who participated in our study wrote: "It is important to build the training capacity of governments rather [than] to rely so heavily on NGOs and international agencies." To ensure that the world is collectively prepared for future epidemics, we need a more collaborative approach that is not guided solely by altruism. The primary purpose of investing in collective readiness is to achieve so-called global health security—a world safe from infectious disease threats. Achieving this goal will involve putting together networks of access to health care, supporting infrastructure and continuous capacity building, and encouraging social mobilization and civil society participation.

ACHIEVING MILITARY PRECISION AND SPEED

People who have responded to an infectious disease epidemic have often likened the environment to that of a war. Far more people around the world have died from infectious diseases than from war, however, and improving health systems has arguably done more to create a stable socioeconomic and political environment than any exercise of military power, and at a lower human and financial cost. Yet, especially in developed countries, response systems are better structured in the military than in public health organizations. To prevent the huge numbers of deaths that could occur from a pandemic, we need a global readiness system with the same level of immediacy, organization, capacity, and competence as is found in developed militaries.

A military-style approach to outbreaks would aid both readiness and response. The success of military operations hinges on readiness in all respects, including planning, personnel, training, equipment, communications, and deployment. Regularly conducted exercises test preparedness and response. This same form of preparation is needed in public health emergency response and must occur before an outbreak even enters the global radar. As Jeremy Farrar, director of the research charity Wellcome Trust, put it, "You don't create an army when you think you will need it. You have an equipped army all the time and hope you'll never have to use it."[8]

Military practices offer other useful models. In principle, military units account for all tools, supplies, and personnel. They have a strict chain of command with established channels of communication. They have a large, dedicated workforce with the clear mission of keeping their country safe. They have a strong sense of "we" rather than "I" in their response to emergencies. They focus on identifying their objective and eliminating the threat. They undertake long deployments. While the CDC deployed staff for periods of about six weeks during the Ebola epidemic, the US Department of Defense uses much longer deployment periods even during volatile situations. It is worth treating outbreaks like other warlike threats and investing in and building capacity for a more sustained response until the threat is eliminated.

There is opportunity, too, for stronger collaboration between public health agencies and militaries of developed countries. The US military is among those that understand the need for aligning incentives for preparedness. The US Army estimates that if a severe infectious disease pandemic were to occur today, the number of US deaths would be nearly double the total number of battlefield deaths that occurred in all of America's wars since the American Revolution.[9] Making readiness for natural and intentional threats a mandate to ensure global health security is another example of aligning incentives. For example, wars of the future may involve biological weapons and engineered microbes, for which both the military and public health responders must be ready. Things will become only more complex and destabilizing when viruses are constructed to override current preventive methods or target one population. Current outbreak culture will only work to exacerbate a devastating situation. While international outbreak response has often included military presence from various host and responding countries, it is worth forging a stronger relationship between the two entities and drawing from both clinical and organizational strengths that each have to provide.

A military-style approach to global health outbreak response is necessary to manage the network of clinical care, data collection, research, and building of infrastructure. In addition, more rigorous governance is needed to ensure transparency and accountability of contracts and missions.

CREATING A CENTRALIZED GOVERNING STRUCTURE

One of the greatest obstacles to effective outbreak response is the lack of a leader who would be responsible for assigning roles among the numerous actors involved. Nor is there a central governing body authorized to establish institutional jurisdictions or hold these institutions accountable. Even the largest agencies that are typically involved in outbreak response lack the structural and organizational capabilities needed to oversee a unified response. In an appeal to reform the WHO, Margaret Chan, the former director-general, acknowledged the agency's failings on Ebola and cited "inadequacies and shortcomings in this organization's administrative, managerial, and technical infrastructures."[10] While reforms within agencies are critical, however, they are not enough. Despite reorganization efforts to overcome bureaucratic structures, agencies like the WHO lack the ability to make swift and sweeping decisions, exert leadership among all other responding agencies, and remain flexible in rapidly changing environments—all essential during public health emergencies.

Effective outbreak response requires communication, coordination, collaboration, and real-time sharing of information across disciplines and with all involved groups, including civil society, national and local governments, nongovernmental regional and international organizations, academic institutions, and the military. A new command center is needed to create guidelines for better harmonization between all players and to encourage collaboration. This centralized governing structure should reward transparency, collaboration, honesty, and accountability. If focused on global infectious disease identification, prevention, and preparedness, it would go a long way toward shifting the focus from response to readiness and eliminating the worst features of outbreak culture.

A centralized command center, on the model of a military command structure, could organize the work of all involved agencies, funneling communication through a single channel and coordinating the response. A system with numerous governing tiers and a detailed reporting structure could help empower local groups to build up health systems in preparation for outbreaks.

The Ebola responders who participated in our survey indicated their wish for a structure of this kind. When asked for suggestions for how to create a collaborative environment and what could be done to improve outbreak response, many respondents recommended a hierarchy within the global health system, with a designated leader to oversee centralized information coming from every stakeholder.

UNIFYING COMMUNICATION SYSTEMS AND APPROACH

Public health emergencies encourage isolationist tendencies, but it is only through unification and collaboration in both attitude and execution that we will create an effective global response system. The global community must not only acknowledge its interconnectedness, but operate with transparent and efficient modes of governance and communication. Conflicting policies and lack of standard protocols, as we have seen, can prevent or delay response to a public health emergency. One health worker who answered our survey wrote, "Epidemics can be stopped much more effectively with adherence to Standards of Procedures on all health procedures. There must also be clear communication and centralized information sharing systems."

Uniform policies among agencies would help create a system that can respond quickly. The more efficiently information is gathered, and the earlier it is disseminated, the better international agencies will be able to understand the scope of the problem and provide the necessary resources to public health and clinical responders. The need for a standardized approach also applies to academic institutions and aid organizations supporting volunteers by providing funding, protection and training, and follow-up physical and mental healthcare. When asked to describe the biggest obstacles while participating in outbreak response, some study respondents wrote that it adds to the confusion when different agencies each supporting their own teams have different protocols. Respondents thought there should be a standardized approach for training, supporting, and supplementing front-line Ebola response workers (across teams and geographic areas).

RESEARCH GOVERNANCE

A unified approach is also needed for the research that is conducted during public health emergencies. While the WHO's International Health Regulations provide a legal structure requiring countries to prevent, detect, and contain disease outbreaks, there is no international framework for incorporating research that is done before and during epidemics.[11]

In December 2015—just after the peak of the Ebola epidemic—the WHO pulled together a list of the pathogens most likely to generate epidemics. Among them, of course, was the Ebola virus. The following May, the agency published a report called *An R&D Blueprint for Action to Prevent Epidemics,* which laid out a plan for promoting research on the most dangerous pathogens.[12] The document outlines a roadmap to fast-track research and development so as to be able to make tests, vaccines, and medicines available during epidemics. There must now be a governing structure that implements integration of research into the system of emergency preparedness and response.

Under a centralized governing system, a task force would be charged with creating partnerships to identify common needs and to conduct the necessary surveillance, detection, and clinical research warranted by the circumstances. Depending on the level of research needed for a particular pathogen, the task force's role might include overseeing trials conducted to determine the origin and trajectory of the pathogen or the effectiveness of particular diagnostics or therapeutics. The task force, made up of people recognized as global leaders in research governance, should be devoid of conflicts of interest so there would be no incentive to prioritize particular agencies, research groups, or individuals.

The task force could make sure that research is undertaken before epidemics start. Usually, new diagnostics and therapeutics are not developed until after an outbreak is well under way, and the product may be introduced only after the epidemic has waned, if at all. This was the case during the 2009 influenza pandemic, when vaccine development ramped up, but a vaccine only became available after the height of the epidemic. And influenza is a well-known threat that researchers have spent years preparing for—it could take much longer to develop a treatment for a less familiar pathogen.

Research on diagnostics and therapeutics conducted during the 2014–2016 Ebola outbreak helped healthcare workers better respond to the outbreak that began in May 2018 in the Democratic Republic of the Congo (DRC). The country's regulatory authorities approved use of the unlicensed experimental Ebola vaccine that had been tested in 2015 in Guinea.[13] Thousands of people in DRC considered to be at highest risk for infection, such as healthcare workers, received the vaccine in an effort to stop Ebola from spreading—a scenario that would not have occurred had the vaccine trials not been undertaken during the earlier outbreak. It is a testament to the importance of working through epidemics, and during the interepidemic period.

Many clinical trials have the same requirements regarding data sharing, patient confidentiality, and informed consent, so the framework for such trials could be prepared ahead of an outbreak, saving time later. If research plans were put in place during the interepidemic period, all parties would know what type of research was going to be conducted, and by whom, during an outbreak. During the actual outbreak, the many actors involved all have their own "vested interests, ideologies, capabilities, mandates, and authorities."[14] But under a centralized research governing model, task force participants would be vetted, chosen, and trained for outbreak research in advance. Aligning incentives under a central leadership would reduce chaos, toxic competition, unnecessary overlap, coveting of data, and restrictions on research.

SECURING FINANCIAL RESOURCES AND OVERSIGHT

Rapid response rooted in uniform policies is crucial for minimizing the economic cost of epidemics. The economic impact relates directly to the speed of national and international response and how quickly the epidemic is contained. The Commission on a Global Health Risk Framework for the Future estimated that $4.5 billion a year is needed to bolster local health systems, enhance global capabilities and coordination, and accelerate research and development.[15] This is a manageable investment, considering that global military spending amounts to nearly $2 trillion.[16]

In addition to funds supplied by governments, a global pandemic preparedness fund would enable faster and more effective outbreak response. Priority must be given to funding at both the regional and international levels to develop diagnostic tools, therapies, and vaccines for relatively rare but potentially devastating diseases between epidemic periods. There must also be a preapproved and ethical mechanism for accelerating development and testing once an epidemic emerges.[17] During the Ebola and other epidemics, scientists experienced firsthand how the time-consuming practices of drafting contracts, conducting meetings, and gaining approval to conduct clinical trials delayed their urgent work.

We need better financial governance to ensure that money allocated to emergencies reaches the intended recipients and appropriately funds the intended outcome. Every dollar spent during an outbreak should be tracked by a centralized system. Outbreaks are more likely to spiral into epidemics in regions with poor infrastructure and governance, inadequate health systems, and rampant corruption. This is all the more reason for such a system. The largest federal funders during public health emergencies should be mandated to disclose and justify every contract. No-bid contracts should be more heavily scrutinized because they can reduce competition for innovation and provide unclear direction and reasons behind the allocation of funds. They can add to the perception of unfair distribution of funds and exacerbate outbreak culture.

PUBLIC HEALTH OVER POLITICS

Global health security requires removing political considerations from decisions about health assistance. Some governments follow a "country-first" approach that, although it may be beneficial to the domestic infrastructure and economy, is detrimental to achieving global health security. If we agree that pathogens are borderless, we must operate a global health system that is indifferent to borders.

Within countries, relying on politicians to set forth a global health agenda has meant forgoing the rigorous scientific analysis and debate necessary to cope with emerging infectious diseases. As we saw during the Ebola epidemic, the decisions that governments and politicians make regarding outbreak response are frequently based on emotion and appeasement of public perception rather than on science.

Developed countries, in particular, have a responsibility to ensure global health security, though it seems that the importance of prioritizing public health over politics has been disregarded despite the lessons learned from recent outbreaks. In May 2018, as part of a list of items aimed at "protecting American taxpayer dollars," President Donald Trump requested that Congress rescind $252 million in "excess funds" allocated to respond to the 2014 Ebola outbreak, on the grounds that the epidemic had been declared over in 2016. The same day, the DRC declared a new Ebola outbreak and requested international aid.[18] The United States has remained mostly silent during this new Ebola outbreak. Earlier in the year, the CDC scaled back its presence in some of the highest-risk countries in anticipation that the Trump administration would likely not extend funding for the agency's Global Health Security Agenda, set to lapse in 2019.[19] The implications for reducing funding and neglecting global health security extend beyond immediate national political gains. They place the United States and the world at increased risk for a disease disaster.

Smaller-scale politics also affects the groups that are directly involved in outbreak response. Scientists and research groups may engage in tactics that contribute to the already toxic environment of a health crisis. Restrictions on research and lack of access to data often lead to redundant work. Local health workers suffer when new responders arrive and implement completely different directives from those that were in place before. One of the most critical lessons learned during the Ebola epidemic is that health workers must constantly adapt to an unpredictable environment while building upon previous knowledge. Leaving politics behind means letting go of possessive behaviors and regarding one's work as continuing that of previous responders. Those programs that were able to adapt quickly to rumors, violence, misunderstandings, and new scientific findings were able to advance much more quickly.

Improving outbreak response does not necessarily require overhauling entire political systems, but rather forming collaborative relationships between public health and political leaders. Public health leaders should reach out to political leaders and large, nonpartisan scientific groups to offer their expert recommendations.

PROTECTING OUR PROTECTORS

Healthcare workers on the front lines during an outbreak are central to early response and containment efforts. Most work tirelessly to save others while risking their own health. Many leave their families behind for days or weeks. Some do not survive. These workers are often stigmatized by their communities and receive little to no insurance or financial compensation for their efforts. Experienced healthcare workers are the most qualified to respond to the next outbreak—yet how can the global health community expect them to participate in future outbreaks if they were not taken care of during the previous one? Effective outbreak response relies on individuals like Liberia's Darlington Jallah to lead early efforts, offer sound clinical judgment based on knowledge of the local traditions, and carry the community through distress. Healthcare providers and others on the front lines need to be protected, cared for, and supported in every way possible. The global community must ensure that they receive timely compensation, insurance, continuing education, and care for their families should they die.

SPEAKING THE SAME LANGUAGE

Effective communication during an epidemic depends on recognizing our shared humanity while respecting our differences. We

Three nurses in training stand in the door of Kenema Government Hospital's nurses' teaching rooms, months before the Ebola outbreak hit. *Photo by Pardis Sabeti/© Sabeti Lab.*

cannot underestimate the importance of understanding the social and cultural complexities of different environments. The way people respond to a disease outbreak is conditioned by their cultural and political history. An important lesson from previous outbreaks is that response teams should include anthropologists, who can understand local knowledge and craft effective tactics for minimizing fear.

Public health messaging needs to be tailored to the people and their environment. Effective campaigns directed to West Africans during the Ebola epidemic looked much different from those directed to Americans. Although the same virus was affecting both regions, perceptions of risk, actual risk, and response to risk were all vastly different. Thus, responders could not rely on existing communication materials recycled from another context, or even a previous epidemic.

Many people in Liberia, Sierra Leone, Guinea, and neighboring countries still do not believe there was ever an Ebola epidemic. But by the time it was over, more people were adhering to safe practices and seeking out healthcare than before the outbreak. Outbreak response needs to be organized around meaningful collaboration from the beginning, in ways that involve local people and diverse knowledge. Early, frequent, and transparent communication on the part of responding organizations is critical to altering the course of an outbreak.

EMPOWERING EVERYONE

Effective intervention requires more than understanding the sociopolitical and cultural environment in which an outbreak emerges. One of the key lessons of the Ebola outbreak was that a top-down approach, whereby large agencies dictate strategies, does not work. Civil societies and communities must be engaged and enabled to take ownership for strengthening health systems within their own communities long before an outbreak. While the work of local healthcare providers is an essential part of outbreak response, others without medical expertise must also be involved. Outbreak control hinges on the collaboration of the entire affected community as an outbreak progresses. A community-oriented model worked well in helping contain some previous epidemics, such as the 2005 Marburg hemorrhagic fever epidemic in Angola, where responding agencies, including the MSF, were initially met with fear and resistance. Within a month, MSF adapted agency procedures to actively involve the community in monitoring and reporting cases and to incorporate local customs and traditions, which helped change the course of the epidemic.[20] It was a lesson learned nearly a decade before the Ebola outbreak in West Africa.

An example of an empowering strategy that was used during the Ebola outbreak is community-based surveillance: designated local volunteers monitored and reported health changes among individuals within their town or village using mobile phones.[21] This program partnered people from communities—in this case in Sierra Leone—with health workers from organizations like the Red Cross to collect data that made early intervention possible. Strengthening community engagement and participation in the interepidemic period makes the battle much easier when an epidemic does arise.

RESPONSIBLE REPORTING OF OUTBREAKS

The media plays a large role in outbreak culture. Media coverage of public health emergencies and epidemics can be misinformed at worst, and inadequate at best. The longer the situation continues, the greater the threat of complacency. News organizations considered the Ebola outbreak to be "breaking news" when a growing number of people were becoming infected with the disease in Africa, and then when a case occurred in the United States. These news stories usually last for only a few days until the news cycle—and thus the public eye—moves on. The epidemic continues, however, even though it is no longer getting airtime on television or a prominent place in newspapers or on news sites. Decision makers in newsrooms determine that the public has lost interest and look for another breaking story to entice their viewers and readers.

This pattern has created an audience of partially informed news consumers who are losing their ability to fully grasp a story in deep context. Journalists and news organizations should be challenged to find engaging ways to offer more informed and comprehensive reporting. Some of this is already happening through storytelling approaches such as "solutions journalism," where narratives focus on how individuals, communities, or systems are working to solve complex issues, or "slow journalism," where the focus is on understanding the ethnography of a region and the people directly affected by the issue. There are dozens of other alternatives to the breaking news cycle. Using these approaches, journalists can tell the stories of key players involved in outbreak response, explain the risk of infection, expose bad practices, and hold the people in power accountable.

CONVEYING KNOWN UNKNOWNS

New and even re-emerging health threats come with uncertainties. Quick and decisive action often has to be taken before all the facts are known; thus, it is imperative to support a culture that promotes unrestricted field research and real-time data sharing. The public, too, demands to be kept informed even when much is unknown. The dilemma for experts is that the public often misconstrues uncertainty as incompetence. Research on the communication of risk suggests that transparent messaging from leadership is critical to easing public fears and establishing consistency in the response among community health centers. Merely stating uncertainty is not enough, according to risk communication expert Peter Sandman. Instead, one must "proclaim" uncertainty—that is, express it "emphatically, dramatically, and repeatedly."[22] It is essential to acknowledge uncertainty where it exists and to justify policies by explaining the scientific information behind decisions based on what is—and is not—known.[23] At present, however, regaining public trust through this method is nearly impossible because many public health decisions are not based on science at all.

PREPARING FOR THE NEXT EPIDEMIC

Multiple international, independent assessment panels were assembled after the 2014–2016 Ebola epidemic, all of which offered similar recommendations about how to improve outbreak prevention and response. Many of the same suggestions had been made after the 1995 Ebola outbreak in Zaire, such as the need for more broad-based international health regulations, a more streamlined information system, and continued and coordinated research.[24] The fact that 20 years had gone by with little change in outbreak response shows the immense gap between knowledge and application. While some improvements have been made since the Ebola epidemic, the global community is in no way ready to contain a pandemic.

Pathogens do not recognize national borders, nor do they recognize political, social, ethnic, or religious differences. The risk of an epidemic of a highly contagious virus, drug-resistant bacterium, or other pathogen spreading across continents remains high. The century ahead brings unprecedented challenges, but also the opportunity both to use new technologies and capacities to combat pathogens and to learn to behave more compassionately toward each other. We as a global community need to recognize the ongoing challenges and implement universal principles and a unified structure to approaching the next epidemic, whenever and wherever it occurs.

EPILOGUE

ON AUGUST 11, 2016, 21 clinicians and scientists from West Africa completed a four-week intensive training course on viral diseases at the Sabeti Lab in Cambridge, Massachusetts. The cohort became the third graduating class of a program called the African Center of Excellence for Genomics of Infectious Diseases (ACEGID). ACEGID was the brainstorm of a May 2014 meeting of the Viral Hemorrhagic Fever Consortium, which included Sheikh Humarr Khan, Pardis Sabeti, and Robert Garry. The group agreed that West Africans should acquire the necessary equipment and education to be able to detect, track, and treat all major forms of hemorrhagic fever.

At the ACEGID symposium called "The Future of Outbreak Response"—which marked the end of the summer course—Khan's successor, Dr. Donald Grant, presented photographs of the new hemorrhagic fever treatment center that was near completion in Kenema. Grant had pledged to carry on the original vision of his two predecessors, Khan and Dr. Aniru Conteh—both of whom died from hemorrhagic fever. "Sometimes colleagues call me just to ask, 'Are you still alive?'" Grant said.

Dr. Donald Grant, the new lead physician of Kenema Government Hospital's Lassa fever program. His two predecessors, Dr. Conteh and Dr. Khan, succumbed to Lassa virus and Ebola virus, respectively. *Photo by Pardis Sabeti/© Sabeti Lab.*

More than four thousand miles away, Khan's body was keeping watch over the Khan Center of Excellence for Viral Hemorrhagic Fever Research. The new U-shaped clinical facility, located at Kenema Government Hospital, contains more than two dozen rooms that will be able to house patients with highly infectious diseases while protecting the staff. It also contains laboratories for conducting research on viral hemorrhagic fevers such as Lassa and Ebola. A memorial just inside the main doors marks Khan's final resting place. As of 2018, the center has yet to open for operation, pending the completion of a few structural details. The center was designed to include all of the protective systems that were lacking during the 2014 Ebola outbreak. For example, there are separate entrances for patients and for healthcare staff, with

The burial and memorial site for Dr. Sheikh Humarr Khan is located in front of the new viral hemorrhagic fever research center. *Photo by Anna Lachenauer/© Sabeti Lab.*

room for the staff to don and doff PPE. Had the center been open during the outbreak, it is likely that many healthcare workers would never have contracted Ebola and died.

Khan envisioned that in Kenema, Africans would lead the effort to establish basic public health capacity and would achieve greater autonomy in conducting research and responding to complex diseases. In the laboratory building, next to the new clinical facility, this work is now going on. Lab technicians are studying blood samples of their colleagues who died from Ebola. Their work, in collaboration with the Sabeti Lab and other research groups, is investigating whether people with particular genetic variants had a higher survival rate. The death of so many in Kenema gives added

The Khan Center of Excellence for Viral Hemorrhagic Fever Research at Kenema Government Hospital. Construction of the building began in 2012 and was completed in 2017. The 48-bed ward has space for quarantine of suspected cases, full care and treatment of acute cases, and supportive care during recovery, with complete separation of suspected and confirmed cases. *Photo by Anna Lachenauer/ © Sabeti Lab.*

purpose to those hoping to change the way local communities detect and respond to outbreaks.

These improvements have changed attitudes toward infectious disease in the region. Clinicians and researchers in Kenema, including those who work directly with Sierra Leone's Ministry of Health and Sanitation, are in a better position to deal with a future outbreak than they were before the Ebola crisis. This is due to the investment in physical infrastructure, better communication at the national level, and increased knowledge of common diseases.

Local scientists now have the ability to collect and analyze samples in real time. The new lab at Kenema houses genetic sequencing

Two laboratory scientists, John Demby Sandi (*front left*) and Mambu Momoh, at Kenema Government Hospital perform diagnostic tests on cases of suspected Lassa fever. *Photo by Kayla Barnes/© Sabeti Lab.*

and rapid diagnostic machines, which enables Africans to understand the pathogens afflicting their community without having to send samples abroad. As of 2017, the center employed one head physician, three lab technicians, a phlebotomist, and dozens of support staff, including nurses and outreach teams, who are able to diagnose and treat Ebola and a number of other diseases. Collaborations with colleagues around the world provide opportunities for improvement in research practice, clinical care, and community health. Researchers and clinicians from other parts of West Africa often visit Kenema to learn from the staff. Scientists from Liberia, for example, travel to Kenema to learn diagnostics techniques and to study the ecology of the rodents that transmit Lassa virus.

In January 2018, another new center broke ground: the ACEGID-led genome research center located on the campus of Redeemer's University in Nigeria.[1] Directed by Cameroonian scientist Christian Happi, the center, which will contain classrooms, offices, and laboratories, will give African researchers the capacity to study and diagnose infectious diseases across the region.

These and other new projects are no small feat. They are giving African clinicians and researchers the confidence and ability to craft their own solutions to the threat of infectious disease outbreaks. They are the realization of a dream for the people's fighter.

NOTES

PROLOGUE: THE PEOPLE'S FIGHTER

Epigraph: Draft email message that Khan intended to send to a WHO medical officer, July 5, 2014. Khan forwarded a copy of the message to Sabeti for comments. It is unclear whether he ever sent the message to the WHO officer.

1. M. Senga, K. Pringle, A. Ramsay, et al., "Factors Underlying Ebola Virus Infection among Health Workers, Kenema, Sierra Leone, 2014–2015," *Clinical Infectious Diseases* 63, no. 4 (2016): 454–459.
2. J. Hammer, "'My Nurses Are Dead, and I Don't Know if I'm Already Infected,'" *Matter,* January 12, 2015, https://medium.com/matter/did-sierra-leones-hero-doctor-have-to-die-1c1de004941e.
3. R. Satter and M. Cheng, "AP Investigation: American Company Bungled Ebola Response," Associated Press, March 7, 2016, https://apnews.com/46328e561bfb44b99b2e6937835be957/ap-investigation-american-company-bungled-ebola-response.
4. T. O'Dempsey, "Failing Dr. Khan," in *The Politics of Fear: Médecins sans Frontières and the West African Ebola Epidemic,* ed. M. Hofman and S. Au (New York: Oxford University Press, 2017).
5. S. H. Khan, telephone conversation with Pardis Sabeti, July 5, 2014.
6. Draft email message from Khan to WHO medical officer, July 5, 2014.
7. Sahid Khan, telephone conversation with Salahi, May 27, 2016.
8. "Lassa Fever," Centers for Disease Control and Prevention, June 2, 2015, https://www.cdc.gov/vhf/lassa/.
9. N. Yozwiak, personal communication with Salahi, November 15, 2016.

10. D. G. Bausch, "The Year that Ebola Virus Took Over West Africa: Missed Opportunities for Prevention," *American Journal of Tropical Medicine and Hygiene* 92, no. 2 (2015): 229–232.
11. P. Sabeti, email to Sheikh Humarr Khan, July 18, 2014.
12. Draft email message from Khan to WHO medical officer, July 5, 2014.
13. R. Garry, Tulane University, telephone communication with Salahi, June 1, 2016.
14. Garry interview, June 1, 2016.
15. R. Preston, "The Ebola Wars," *New Yorker,* October 27, 2014.
16. Preston, "Ebola Wars"; X. Qiu, G. Wong, J. Audet, et al., "Reversion of Advanced Ebola Virus Disease in Nonhuman Primates with ZMapp," *Nature* 514 (2014): 47–53.
17. Garry interview, June 1, 2016.
18. O'Dempsey, "Failing Dr. Khan."
19. O'Dempsey, "Failing Dr. Khan"; A. Pollack, "Opting against Ebola Drug for Ill African Doctor," *New York Times,* August 13, 2014, A1.
20. This section based largely on S. Jalloh interview with Salahi.
21. O'Dempsey, "Failing Dr. Khan."

1. SETTING FOR DISASTER

1. C. E. Coltart, B. Lindsey, I. Ghinai, et al., "The Ebola Outbreak, 2013–2016: Old Lessons for New Epidemics," *Philosophical Transactions of the Royal Society of London, B, Biological Sciences* 372, no. 1721 (2017): 20160297; J. J. Farrar and P. Piot, "The Ebola Emergency—Immediate Action, Ongoing Strategy," *New England Journal of Medicine* 371, no. 16 (2014): 1545–1546.
2. Transparency International ranks Guinea, where the outbreak likely started, 145 out of 175 countries on its corruption scale (the higher the number, the worse the offender). Sierra Leone ranks 119, and Liberia 94. Transparency International, "Corruption Perceptions Index 2014: Results," https://www.transparency.org/cpi2014/results.
3. "Peace Agreement between the Government of the Republic of Sierra Leone and the Revolutionary United Front of Sierra Leone, signed at Abidjan on 30 November 1996," United States Institute of Peace,

Peace Agreements Digital Collection, https://www.usip.org/sites/default/files/file/resources/collections/peace_agreements/sierra_leone_11301996.pdf.

4. Information and quotations in this section from Sahid Khan, interview with Salahi, May 27, 2016.
5. Mariama Lahai, interview with Salahi, January 30, 2017.
6. "Sahr and Nyumah," Stories of Peace, Fambul Tok International, http://www.fambultok.org/what-is-fambul-tok/stories/sahr-and-nyumah.
7. M. Mayhew, "Demand for Mental Health Services Surges in Liberia and Sierra Leone," World of Opportunity, World Bank via *Medium,* April 12, 2016, https://medium.com/world-of-opportunity/demand-for-mental-health-services-surges-in-liberia-and-sierra-leone-a59fe19f51dc.
8. Ryan Lenora Brown, "Ebola, War . . . but Just Two Psychiatrists to Deal with a Nation's Trauma," *Guardian,* January 20, 2017, https://www.theguardian.com/world/2017/jan/20/sierra-leone-war-ebola-africa-psychiatric-care.
9. "What Is Fambul Tok?" Fambul Tok International, http://www.fambultok.org/what-is-fambul-tok; L. S. Graybill, "Traditional Practices and Reconciliation in Sierra Leone: The Effectiveness of Fambul Tok," *Conflict Trends* 3 (2010): 41–47.
10. On some of the drawbacks to the reconciliation process, see J. Cilliers, O. Dube, and B. Siddiqi, "Reconciling after Civil Conflict Increases Social Capital but Decreases Individual Well-Being," *Science* 352 no. 6287 (2016): 787–794.
11. S. Baize, D. Pannetier, L. Oestereich, et al., "Emergence of Zaire Ebola Virus Disease in Guinea," *New England Journal of Medicine* 371, no. 5 (2014): 1418–1425; S. Briand, E. Bertherat, P. Cox, et al., "The International Ebola Emergency," *New England Journal of Medicine,* 371, no. 13 (2014): 1180–1183; WHO, "Ground Zero in Guinea: The Ebola Outbreak Smoulders—Undetected—for More than 3 Months," undated, http://www.who.int/csr/disease/ebola/ebola-6-months/guinea/en/.
12. C. Zimmer, *A Planet of Viruses,* 2nd ed. (Chicago: University of Chicago Press, 2015).

13. N. Wauquier, J. Banguar, L. Moses, et al., "Understanding the Emergence of Ebola Virus Disease in Sierra Leone: Stalking the Virus in the Threatening Wake of Emergence," *PLOS Currents Outbreaks,* April 20, 2015, https://www.ncbi.nlm.nih.gov/pmc/articles/PMC4423925; K. Sack, S. Fink, P. Belluck, and A. Nossiter, "How Ebola Roared Back," *New York Times,* December 29, 2014, https://www.nytimes.com/2014/12/30/health/how-ebola-roared-back.html?_r=0; C. W. Dugger, "How *Times* Reporters Unraveled the Ebola Epidemic," *New York Times,* January 2, 2015, https://www.nytimes.com/times-insider/2015/01/02/how-times-reporters-unraveled-the-ebola-epidemic/.
14. A. Goba, S. H. Khan, M. Fonnie, et al., "An Outbreak of Ebola Virus Disease in the Lassa Fever Zone," *Journal of Infectious Diseases* 214, suppl. 3 (2016): S110–S121.
15. S. Gire, A. Goba, K. G. Andersen, et al., "Genomic Surveillance Elucidates Ebola Virus Origin and Transmission during the 2014 Outbreak," *Science* 345, no. 6202 (2014): 1369–1372.
16. D. Park, G. Dudas, S. Wohl, et al., "Ebola Virus Epidemiology, Transmission, and Evolution during Seven Months in Sierra Leone," *Cell* 161, no. 7 (2015): 1516–1526.
17. Sack, Fink, Belluck, and Nossiter, "How Ebola Roared Back."
18. R. A. Urbanowicz, P. C. McClure, A. Sakuntabhai, et al., "Human Adaptation of Ebola Virus during the West African Outbreak," *Cell* 167, no. 4 (2016): 1079–1087.
19. W. E. Diehl, A. E. Lin, N. D. Grubaugh, et al., "Ebola Virus Glycoprotein with Increased Infectivity Dominated the 2013–2016 Epidemic," *Cell* 167, no. 4 (2016): 1088–1098.
20. A. P. Galvani and R. M. May, "Epidemiology: Dimensions of Superspreading," *Nature* 438, no. 7066 (2005): 293–295; A. Kucharski and C. Althaus, "The Role of Superspreading in Middle East Respiratory Syndrome Coronavirus (MERS-CoV) Transmission," *Eurosurveillance* 20, no. 25 (2015): 14–18.
21. M. S. Y. Lau, B. D. Dalziel, S. Funk, et al., "Spatial and Temporal Dynamics of Superspreading Events in the 2014–2015 West Africa Ebola Epidemic," *Proceedings of the National Academy of Sciences* 114, no. 9 (2017): 2337–2342.

22. WHO Ebola Response Team, "Ebola Virus Disease in West Africa—The First 9 Months of Epidemic and Forward Projections," *New England Journal of Medicine* 371, no. 16 (2014): 1481–1495.
23. J. S. Schieffelin, J. G. Shaffer, A. Goba, et al., "Clinical Illness and Outcomes in Patients with Ebola in Sierra Leone," *New England Journal of Medicine* 371, no. 22 (2014): 2092–2100; X. Liu, E. Speranza, C. Muñoz-Fontela, et al., "Transcriptomic Signatures Differentiate Survival from Fatal Outcomes in Humans Infected with Ebola Virus," *Genome Biology* 18, no. 1 (2017), https://doi.org/10.1186/s13059-016-1137-3.
24. "What Are the Risks of Ebola Recurring?" BBC News, October 21, 2015, http://www.bbc.com/news/health-34485116.
25. A. A. Adalja, "Sanctuary Sites: What Lies behind Ebola Eye Infections, Sexual Transmission, and Relapses," *Health Security* 13, no. 6 (2015): 396–398; K. G. Barnes, J. Kindrachuk, A. E. Lin, et al., "Evidence of Ebola Replication and High Concentration in Semen of a Patient during Recovery," *Clinical Infectious Diseases* 65, no. 8 (2017): 1400–1403.
26. J. R. Glynn, "Age-specific Incidence of Ebola Virus Disease," *Lancet* 386, no. 9992 (2015): 432.
27. I. Trehan, T. Kelly, R. H. Marsh, P. M. George, and C. W. Callahan, "Moving towards a More Aggressive and Comprehensive Model of Care for Children with Ebola," *Journal of Pediatrics* 170 (2016): 28–33.
28. National Academy of Sciences, Engineering, and Medicine, *The Ebola Epidemic in West Africa: Proceedings of a Workshop* (Washington, DC: National Academies Press, 2016), chap. 2, "The Outbreak," available at https://www.nap.edu/catalog/23653/the-ebola-epidemic-in-west-africa-proceedings-of-a-workshop.
29. M. Osterholm, telephone interview with Salahi, January 24, 2018.
30. USAID / CDC, "West Africa—Ebola Outbreak," Fact Sheet #12, fiscal year 2016, September 30, 2016, https://www.usaid.gov/sites/default/files/documents/1866/west_africa_ebola_fs12_09-30-2016.pdf; L.-Q. Fang, Y. Yang, J.-F. Jiang, et al., "Transmission Dynamics of Ebola Virus Disease and Intervention Effectiveness in Sierra Leone," *Proceedings of the National Academy of Sciences* 113, no. 16 (2016): 4488–4493.

31. WHO Ebola Response Team, "Ebola Virus Disease in West Africa."
32. Farrar and Piot, "The Ebola Emergency."
33. Farrar and Piot, "The Ebola Emergency."

2. THE CRUCIBLE OF OUTBREAK RESPONSE

1. See J. Hammer, "'My Nurses Are Dead, and I Don't Know if I'm Already Infected,'" *Matter,* January 12, 2015, https://medium.com/matter/did-sierra-leones-hero-doctor-have-to-die-1c1de004941e.; R. Preston, "The Ebola Wars," *New Yorker,* October 27, 2014; J. Cohen, "Ebola Survivor II, Nancy Writebol: 'We Just Don't Even Have a Clue What Happened,'" *Science* News, October 2, 2014, http://www.sciencemag.org/news/2014/10/ebola-survivor-ii-nancy-writebol-we-just-dont-even-have-clue-what-happened. Details of this episode were confirmed independently.
2. "Report of the Independent Panel on the U.S. Department of Health and Human Services (HHS) Ebola Response," June 2016, https://www.phe.gov/Preparedness/responders/ebola/EbolaResponseReport/Documents/ebola-panel.pdf.
3. R. Garry, telephone interview with Salahi, June 1, 2016.
4. T. O'Dempsey, "Failing Dr. Khan," in *The Politics of Fear: Médecins sans Frontières and the West African Ebola Epidemic,* ed. M. Hofman and S. Au (New York: Oxford University Press, 2017), 175–186.
5. R. Quigley, vice president, International SOS, telephone conversation with Salahi, July 11, 2016.
6. S. Jalloh, Skype interview with Salahi, July 1, 2016.
7. World Health Organization, "Experimental Therapies: Growing Interest in the Use of Whole Blood or Plasma from Recovered Patients (Convalescent Therapies)," September 26, 2014, http://www.who.int/mediacentre/news/ebola/26-september-2014/en/.
8. J. Farrar, telephone conversation with Salahi, April 28, 2017.
9. J. Farrar, telephone conversation with Salahi, April 28, 2017.
10. P. Hitchcock, A. Chamberlain, M. Van Wagoner, et al., "Challenges to Global Surveillance and Response to Infectious Disease Outbreaks of International Importance," *Biosecurity and Bioterrorism: Biodefense Strategy, Practice, and Science* 5, no. 3 (2007): 206–227.

11. S. Davis, telephone conversation with Salahi, September 16, 2016.
12. C. E. Rosenberg, *Explaining Epidemics and Other Studies in the History of Medicine* (Cambridge: Cambridge University Press, 1992).
13. World Health Organization, *International Health Regulations (1969)*, 3rd annotated ed. (Geneva: World Health Organization, 1983), http://www.who.int/csr/ihr/ihr1969.pdf.
14. M. Hume, "Federal Agency Accused of Intimidation over Salmon Disease," *Globe and Mail* (Toronto), December 16, 2011, updated online March 26, 2017, http://www.theglobeandmail.com/news/british-columbia/federal-agency-accused-of-intimidation-over-salmon-disease/article554785/.
15. R. D. Smith, "Responding to Global Infectious Disease Outbreaks: Lessons from SARS on the Role of Risk Perception, Communication and Management," *Social Science and Medicine* 63, no. 12 (2006): 3113–3123.
16. "China Health Minister Says SARS under Control," attributed to Reuters 2003, Rense.com Alt News, http://www.rense.com/general36/control.htm.
17. Hitchcock et al., "Challenges to Global Surveillance"; I. K. Damon, C. E. Roth, and V. Chowdhary, "Discovery of Monkeypox in Sudan," letter to the editor, *New England Journal of Medicine* 355 (2006): 962–963.
18. C. E. Rosenberg, *The Cholera Years: The United States in 1832, 1849, and 1866* (Chicago: University of Chicago Press, 1987).
19. P. Washer, "Representations of SARS in the British Newspapers," *Social Science and Medicine* 59 (2004): 2561–2571.
20. U.S. Department of Health and Human Services, "A Timeline of HIV / AIDS," AIDS.gov, https://www.hiv.gov/sites/default/files/aidsgov-timeline.pdf.
21. Mark Joseph Stern, "Listen to Reagan's Press Secretary Laugh about Gay People Dying of AIDS," Slate, Outward, December 1, 2015, http://www.slate.com/blogs/outward/2015/12/01/reagan_press_secretary_laughs_about_gay_people_dying_of_aids.html; A. M. Brandt, "AIDS: From Social History to Social Policy," *Journal of Law, Medicine and Ethics* 14, no. 5–6 (1986): 231–242.
22. Centers for Disease Control and Prevention, "Opportunistic Infections and Kaposi's Sarcoma among Haitians in the United States,"

Morbidity and Mortality Weekly Report 31, no. 26 (July 9, 1982): 353–354, 360–361.

23. S. Davis, telephone conversation with Salahi, September 16, 2016.
24. A. Orkin, "Boycott Casts Shadow over San Francisco AIDS Conference," *Canadian Medical Association Journal* 142, no. 12 (1990): 1411–1413, https://www.ncbi.nlm.nih.gov/pmc/articles/PMC1451963/pdf/cmaj00217-0061.pdf.
25. S. J. Hoffman and S. L. Silverberg, "Delays in Global Disease Outbreak Responses: Lessons from H1N1, Ebola, and Zika," *American Journal of Public Health* 108, no. 3 (2018): 329–333.
26. M. A. Connolly, M. Gayer, M. J. Ryan, et al., "Communicable Diseases in Complex Emergencies: Impact and Challenges," *Lancet* 364, no. 9449 (2004): 1974–1983.
27. R. Holbrooke and L. Garrett, "'Sovereignty' That Risks Global Health," *Washington Post,* August 10, 2008.
28. P. Piot, telephone conversation with Sabeti and Salahi, April 6, 2017.
29. K. Kupferschmidt, "As Outbreak Continues, Confusion Reigns over Virus Patents," News from *Science,* May 28, 2013, http://www.sciencemag.org/news/2013/05/outbreak-continues-confusion-reigns-over-virus-patents.
30. Smith, "Responding to Global Infectious Disease Outbreaks."
31. GHRF Commission (Commission on a Global Health Risk Framework for the Future), "The Case for Investing in Pandemic Preparedness," in *The Neglected Dimension of Global Security: A Framework to Counter Infectious Disease Crises,* National Academy of Medicine (Washington, DC: National Academies Press, 2016), chap. 2, https://www.ncbi.nlm.nih.gov/books/NBK368391/. See also J. W. Lee and W. J. McKibbin, "Estimating the Global Economic Costs of SARS," in Institute of Medicine, *Learning from SARS: Preparing for the Next Disease Outbreak: Workshop Summary* (Washington, DC: National Academies Press, 2004), 92–109.
32. Smith, "Responding to Global Infectious Disease Outbreaks."
33. "Haiti Violence Hampering Cholera Response, UN and Partners Warn," UN News, November 17, 2010, http://www.un.org/apps/news/story.asp?NewsID=36787#.WW40QtPyuu4.
34. J. M. Katz, "U.N. Admits Role in Cholera Epidemic in Haiti," *New York Times,* August 17, 2016.

35. Smith, "Responding to Global Infectious Disease Outbreaks," 3117.
36. S. Gire, telephone conversation with Salahi, December 30, 2016.
37. P. Piot, telephone conversation with Sabeti and Salahi, April 6, 2017.
38. R. Bazell, "Dispute behind Nobel Prize for HIV Research," NBC News, Second Opinion online, October 6, 2008, http://www.nbcnews.com/id/27049812/ns/health-second_opinion/t/dispute-behind-nobel-prize-hiv-research/#.WO5ApFPyv-Y.
39. V. A. Harden, "The NIH and Biomedical Research on AIDS," in *AIDS and the Public Debate: Historical and Contemporary Perspectives,* ed. C. Hannaway, V. A. Harden, and J. Parascandola (Washington, DC: IOS Press, 1995), 30–46.
40. B. Conton, "Build the Ebola Database in Africa," *Nature* 551, no. 7679 (2017): 143.
41. J. Fair, telephone conversation with Salahi, July 7, 2017.
42. Congressional seminar on the Ebola outbreak in West Africa, C-Span, September 24, 2014, user-created clip, https://www.c-span.org/video/?c4509906/joseph-fair-sierra-leone-adviser-health-minister-joins-address-senate-hearing.
43. N. G. Evans, T. C. Smith, and M. S. Majumder, *Ebola's Message: Public Health and Medicine in the Twenty-First Century* (Cambridge, MA: MIT Press, 2016).
44. A. Cooper, correspondent, "60 Minutes Investigates Medical Gear Sold during Ebola Crisis," aired May 1, 2016, CBS News, script available at http://www.cbsnews.com/news/60-minutes-investigates-medical-gear-sold-during-ebola-crisis/.
45. "Jury Hits Kimberly-Clark and Halyard Health with $454 Million Fraud Verdict over Sale of Defective Medical Devices Announces Eagan Avenatti, LLP," April 10, 2017, Business Wire, https://www.businesswire.com/news/home/20170410005908/en/Jury-Hits-Kimberly-Clark-Halyard-Health-454-Million. This amount was reduced in 2018 to $1.6 million for Halyard Health and $24.4 million for Kimberly Clark. "Halyard Health Dodges One Suit, Sees Damages Slashed in Another," April 2, 2018, MassDevice, https://www.massdevice.com/halyard-health-dodges-one-suit-sees-damages-slashed-in-another/.

3. THE CASE FOR COLLABORATION

1. Material in this section is from N. Yozwiak, interviews with Salahi on November 15, 2016, and July 21, 2017, and S. K. Gire, interview with Salahi on December 30, 2016.
2. M. Cheng, "AP Investigation: Bungling by UN Agency Hurt Ebola Response," Associated Press, September 21, 2015, https://apnews.com/3ba4599fdd754cd28b93a31b7345ca8b; R. Satter, "AP Investigation: American Company Bungled Ebola Response," Associated Press, March 7, 2016, https://apnews.com/46328e561bfb44b99b2e6937835be957/ap-investigation-american-company-bungled-ebola-response.
3. Satter, "AP Investigation."
4. T. O'Dempsey, "Failing Dr. Khan," in *The Politics of Fear: Médecins sans Frontières and the West African Ebola Epidemic,* ed. M. Hofman and S. Au (New York: Oxford University Press, 2017).
5. This section based on Salahi's phone conversations with Humarr Khan's brother Sahid Khan (January 15, 2017) and sister Mariama Lahai (January 30, 2017).
6. S. K. Gire, A. Goba, K. G. Andersen, et al., "Genomic Surveillance Elucidates Ebola Virus Origin and Transmission during the 2014 Outbreak," *Science* 345, no. 6202 (2014): 1369–1372.
7. *Zaire ebolavirus* genome sequencing, BioProject, NCBI, US National Library of Medicine, National Institutes of Health, http://www.ncbi.nlm.nih.gov/bioproject/PRJNA257197.
8. "Sierra Leone News: Hero in War against Ebola May be Smiling in His Grave," *Awareness Times,* accessed August 29, 2014, http://news.sl/drwebsite/exec/view.cgi?archive=10&num=26093.
9. "Sierra Leone Dismisses Health Minister over Handling of Ebola," Reuters, August 29, 2014, https://www.reuters.com/article/us-health-ebola-leone/sierra-leone-dismisses-health-minister-over-handling-of-ebola-idUSKBN0GT2DO20140829.
10. Email message from Monty Jones, special adviser to the president and ambassador-at-large, Sierra Leone, to Sylvia Blyden, August 28, 2014.
11. Gire et al., "Genomic Surveillance Elucidates Ebola Virus Origin."
12. A. Cori, C. A. Donnelly, I. Dorigatti, et. al., "Key Data for Outbreak Evaluation: Building on the Ebola Experience," *Philosophical Transactions of the Royal Society B* 372, no. 1721 (2017): 20160371.

13. World Health Organization, Ebola Situation Report, September 30, 2015, http://apps.who.int/ebola/current-situation/ebola-situation-report-30-september-2015.
14. A. Maxmen, "Huge Ebola Data Site Takes Shape," *Nature* 549, no. 7670 (2017): 15.
15. Quoted in A. Maxmen, "Massive Ebola Data Site Planned to Combat Outbreaks," *Nature* News online, September 4, 2017, https://www.nature.com/news/massive-ebola-data-site-planned-to-combat-outbreaks-1.22545.
16. A. K. Igonoh, "My Experience as an Ebola Patient," *American Journal of Tropical Medicine and Hygiene* 92, no. 2 (2015): 221–222.
17. Ebola Vaccine Team B, "Completing the Development of Ebola Vaccines: Current Status, Remaining Challenges, and Recommendations," January 2017, Center for Infectious Disease Research and Policy, University of Minnesota, http://www.cidrap.umn.edu/sites/default/files/public/downloads/ebola_team_b_report_3-011717-final_0.pdf.
18. T. Lang, "Ebola: Embed Research in Outbreak Response," *Nature* 524, no. 7563 (2015): 29–31.
19. Quoted in K. Kelland, "MERS, Ebola, Bird Flu: Science's Big Missed Opportunities," Reuters, October 26, 2015, http://www.reuters.com/article/us-health-epidemic-research-insight-idUSKCN0SK0P020151026.

4. THE WAVERING RESPONSE

1. N. Bhadelia, telephone conversation with Salahi, March 28, 2016.
2. Biography of Nahid Bhadelia, Boston University Medical Center, http://www.bumc.bu.edu/id/faculty/bhadelia-nahid-m-d-m-a/.
3. E. Sterk, "Filovirus Haemorrhagic Fever Guideline," Médecins sans Frontières, 2008, p. 70, available at https://www.oregonpa.org/resources/2014CME/Speaker%20Presentations/Ebola%20Guideline%20-%20Crawford.pdf.
4. K. Sack, S. Fink, P. Belluck, and A. Nossiter, "How Ebola Roared Back," *New York Times,* December 29, 2014.
5. Sack et al., "How Ebola Roared Back."
6. C. Freeman, "Guinea and Sierra Leone Tried to Cover Up Ebola Crisis, Says Médecins sans Frontières," *The Telegraph,* March 23, 2015.

7. S. J. Hoffman and S. L. Silverberg, "Delays in Global Disease Outbreak Responses: Lessons from H1N1, Ebola, and Zika," *American Journal of Public Health* 108, no. 3 (2018): 329–333.
8. A. J. Kucharski, A. Camacho, S. Flasche, et al., "Measuring the Impact of Ebola Control Measures in Sierra Leone," *Proceedings of the National Academy of Sciences* 112, no. 46 (2015): 14366–14371.
9. Médecins sans Frontières, "Ebola in West Africa: Epidemic Requires Massive Deployment of Resources," MSF website, June 21, 2014, http://www.msf.org/en/article/ebola-west-africa-epidemic-requires-massive-deployment-resources.
10. Médecins sans Frontières, "Pushed to the Limit and Beyond: A Year into the Largest Ever Ebola Outbreak," March 23, 2015, report available at http://www.msf.org/sites/msf.org/files/msf1yearebolareport_en_230315.pdf.
11. E. Buchanan, "Ebola Crisis: MSF and WHO Trade Accusations over Epidemic Response," *International Business Times,* April 9, 2015, http://www.ibtimes.co.uk/ebola-crisis-msf-who-trade-accusations-over-epidemic-response-1493199.
12. Freeman, "Guinea and Sierra Leone."
13. R. Garry, telephone conversation with Salahi and Sabeti, February 9, 2018; email messages from MSF employee to Garry, March 22 and 24, 2015.
14. World Health Organization, "Statement on the 1st Meeting of the IHR Emergency Committee on the 2014 Ebola Outbreak in West Africa," August 8, 2014, http://www.who.int/mediacentre/news/statements/2014/ebola-20140808/en/.
15. M. Cheng and R. Satter, "Emails Show the World Health Organization Intentionally Delayed Calling Ebola a Public Health Emergency," Business Insider website, March 20, 2015, http://www.businessinsider.com/report-the-world-health-organization-resisted-declaring-ebola-an-international-emergency-for-economic-reasons-2015-3.
16. World Health Organization, "Statement on the 1st Meeting of the IHR Emergency Committee on the 2014 Ebola Outbreak in West Africa."
17. World Health Organization, International Health Regulations, http://www.who.int/topics/international_health_regulations/en/.
18. S. Moon, J. Leigh, L. Woskie, et al., "Post-Ebola Reforms: Ample Analysis, Inadequate Action," *BMJ* 356 (2017): j280.

19. J. Yang, "What Went Wrong in Response to the Ebola Crisis?" *Star* (Toronto), October 17, 2014, https://www.thestar.com/news/world/2014/10/17/what_went_wrong_in_response_to_the_ebola_crisis.html.
20. M. Hussain, "Ebola Response of MSF and 'Boiling Frog' WHO under Scrutiny," Reuters, August 12, 2014, http://www.reuters.com/article/us-foundation-health-ebola-response-idUSKBN0GL1TT20140821.
21. Sack et al., "How Ebola Roared Back."
22. World Health Organization, "Successful Ebola Responses in Nigeria, Senegal and Mali," January 2015, http://www.who.int/csr/disease/ebola/one-year-report/nigeria/en/.
23. Quoted in Hussain, "Ebola Response of MSF."
24. Médecins sans Frontières, "Ebola in West Africa."
25. Médecins sans Frontières, "Financial Accountability: Ebola Response in 2014," March 23, 2015, http://www.doctorswithoutborders.org/article/financial-accountability-ebola-response-2014.
26. S. Davis, "Ebola: From Emergency to Recovery" (lecture, Regis College, Weston, MA, April 20, 2016).
27. P. Farmer, "Surviving Ebola and Reflections on Those Who Did Not" (lecture, Harvard Medical School, Boston, MA, September 23, 2016).
28. T. E. C. Jones-Konneh, A. Murakami, H. Sasaki, and S. Egawa, "Intensive Education of Health Care Workers Improves the Outcome of Ebola Virus Disease: Lessons Learned from the 2014 Outbreak in Sierra Leone," *Tohoku Journal of Experimental Medicine* 243, no. 2 (2017): 101–105.
29. J. Liu, "MSF President's Remarks to the UN Special Briefing on Ebola," Geneva, September 16, 2014, available at Médecins sans Frontières website, https://www.msf.org/msf-presidents-remarks-un-special-briefing-ebola; L. Westcott, "Obama to Send 3,000 Military Personnel to Fight Ebola," *Newsweek,* September 16, 2014, http://www.newsweek.com/obama-send-3000-military-personnel-fight-ebola-270919.
30. M. Chan, "WHO Director-General Addresses UN Security Council on Ebola," New York, September 18, 2014, http://www.who.int/dg/speeches/2014/security-council-ebola/en/.
31. United Nations, "With Spread of Ebola Outpacing Response, Security Council Adopts Resolution 2177 (2014) Urging Immediate Action, End to Isolation of Affected States," Security Council

7268th meeting, September 18, 2014, https://www.un.org/press/en/2014/sc11566.doc.htm.

32. World Health Organization, "Key Events in the WHO Response to the Ebola Outbreak," January 2015, http://www.who.int/csr/disease/ebola/one-year-report/who-response/en/.

33. K. Rawlinson, "Ebola Claims Life of Third Doctor in Sierra Leone," *Guardian*, August 27, 2014, https://www.theguardian.com/society/2014/aug/27/ebola-third-doctor-dies-sierra-leone.

34. M. Chan, Address to the Seventieth World Health Assembly, World Health Organization, May 22, 2017, http://www.who.int/dg/speeches/2017/address-seventieth-assembly/en/; M. Chan, "My Decade Leading the WHO: Dirty Fights and Steps toward Universal Coverage," STAT news, June 20, 2017, https://www.statnews.com/2017/06/20/margaret-chan-who-director-general/.

35. N. Yaffe, "Will State Inaction at UN Imperil Haiti Cholera Response?" International Peace Institute, Global Observatory, April 4, 2017, https://theglobalobservatory.org/2017/04/cholera-haiti-minustah-peacekeeping/.

36. "Ban Ki-moon Apologizes for Cholera Outbreak in Haiti," UN Headquarters, New York, December 1, 2016, RUPTLY TV, https://www.youtube.com/watch?v=J5AYihbNtM0.

37. K. Mangan, "Universities Curtail Health Experts' Efforts to Work on Ebola in West Africa," *Chronicle of Higher Education* 61, no. 9 (October 21, 2014), http://www.chronicle.com/article/Universities-Curtail-Health/149543/.

38. N. Aizenman, "Ebola Volunteers Are Needed—But Signing On Isn't Easy," *All Things Considered,* NPR, October 14, 2014, transcript at http://www.npr.org/sections/goatsandsoda/2014/10/14/356144079/ebola-volunteers-are-needed-but-signing-on-isnt-easy.

39. J. L. Anderson, "Cuba's Ebola Diplomacy," *New Yorker,* November 4, 2014.

40. L.-Q. Fang, Y. Yang, J.-F. Jiang, et al., "Transmission Dynamics of Ebola Virus Disease and Intervention Effectiveness in Sierra Leone," *Proceedings of the National Academy of Sciences* 113, no. 16 (2016): 4488–4493.

41. I. Kholikov and K. Sazonova, "The Ebola Response Team Deployment in the Republic of Guinea," in D. Messelken and D. Winkler,

eds., *Ethical Challenges for Military Health Care Personnel: Dealing with Epidemics* (New York: Routledge, 2017).

42. Davis, "Ebola: From Emergency to Recovery."
43. N. Bhadelia, "Health Security and Emerging Infections" (lecture, Fletcher Ideas Exchange, Fletcher School of Diplomacy, Tufts University, Medford, MA, April 14, 2016), available at https://www.youtube.com/watch?v=yY0y0CuK1Q0.
44. A. E. Semper, M. J. Broadhurst, J. Richards, et al., "Performance of the GeneXpert Ebola Assay for Diagnosis of Ebola Virus Disease in Sierra Leone: A Field Evaluation Study," *PLOS Medicine* 13, no. 3 (2016): e1001980.
45. Bhadelia, "Health Security and Emerging Infections."
46. Bhadelia, "Health Security and Emerging Infections."

5. DISTRUST IN A CULTURE OF COMPASSION

Epigraph: E. S. Kollie, B. J. Winslow, P. Pothier, and D. Gaede, "Deciding to Work during the Ebola Outbreak: The Voices and Experiences of Nurses and Midwives in Liberia," *International Journal of Africa Nursing Sciences* 7 (2017): 75–81.

1. M. Cheng, R. Satter, and K. Larson, "AP Investigation: Bungling by UN Agency Hurt Ebola Response," *Chicago Tribune*, September 20, 2015, http://www.chicagotribune.com/news/nationworld/ct-bungled-ebola-respone-20150920-story.html.
2. U. Fofana, "Ebola Center in Sierra Leone under Guard after Protest March," Reuters, July 26, 2014, https://www.reuters.com/article/us-health-ebola-africa/ebola-center-in-sierra-leona-under-guard-after-protest-march-idUSKBN0FV0NL20140726; U. Fofana and J. H. Giahyue, "Health Workers Strike at Major Ebola Clinic in Sierra Leone," *Chicago Tribune,* August 30, 2014, http://www.chicagotribune.com/lifestyles/health/chi-health-workers-strike-ebola-clinic-20140830-story.html.
3. World Health Organization, "Ebola Virus Disease Outbreak—West Africa," September 4, 2014, http://www.who.int/csr/don/2014_09_04_ebola/en/.
4. N. Bhadelia, "Is Stigma an Invisible Killer?" TEDxNatick, March 28, 2016, https://www.youtube.com/watch?v=DR8GOoE91Lk.

5. World Health Organization, "Busting the Myths about Ebola Is Crucial to Stop the Transmission of the Disease in Guinea," April 2014, http://www.who.int/features/2014/ebola-myths/en/; personal communication to Salahi from responders in Guinea; "Ebola Myths: Sierra Leonean DJ Tackles Rumours and Lies over the Airwaves," *Guardian,* October 9, 2014, https://www.theguardian.com/global-development/2014/oct/09/ebola-myths-sierra-leone-dj-tackles-rumours; A. K. Igonoh, "My Experience as an Ebola Patient," *American Journal of Tropical Medicine and Hygiene* 92, no. 2 (2015): 221–222.
6. Centers for Disease Control and Prevention, "Number of Cases and Deaths in Guinea, Liberia, and Sierra Leone during the 2014–2016 West Africa Ebola Outbreak," https://www.cdc.gov/vhf/ebola/history/2014-2016-outbreak/case-counts.html; H. Epstein, "Ebola in Liberia: An Epidemic of Rumors," *New York Review of Books* 61, no. 20 (December 18, 2014), http://www.nybooks.com/articles/2014/12/18/ebola-liberia-epidemic-rumors/; D. Jallah, Skype with Salahi, March 24, 2016.
7. D. Jallah, Skype with Salahi, March 24, 2016.
8. "Ebola Memorial Cemetery Dedicated in Liberia," Samaritan's Purse website, January 19, 2016, https://www.samaritanspurse.org/article/ebola-memorial-cemetery-dedicated-in-liberia/.
9. "Effort Liberia Foya Medical Mission," Effort Baptist Church, EffortMedia, February 28, 2010, https://www.youtube.com/watch?v=j4VJRN9xfKg.
10. Centers for Disease Control and Prevention, "Evidence for a Decrease in Transmission of Ebola Virus—Lofa County, Liberia, June 8–November 1, 2014," *Morbidity and Mortality Weekly Report* 63 (November 14, 2014): 1–5, https://www.cdc.gov/mmwr/preview/mmwrhtml/mm63e1114a1.htm.
11. Health Communication Capacity Collaborative. "Social Mobilization Lessons Learned: The Ebola Response in Liberia," Johns Hopkins Center for Communication Programs, February 2017.
12. Centers for Disease Control and Prevention, "Evidence for a Decrease in Transmission."
13. Jallah argued that the girl should be kept for a couple of additional days for observation. Jallah email to Salahi, March 24, 2016.
14. Sources for the story of Pauline are D. Wilson, phone conversations with Salahi, December 17, 2015, April 13, 2016, and April 20, 2016, and D. Jallah, Skype with Salahi, March 24, 2016.

15. Epstein, "Ebola in Liberia."
16. C. N. Adichie, "The Danger of a Single Story," TED Global Conference, July 2009, https://www.ted.com/talks/chimamanda_adichie_the_danger_of_a_single_story#t-590676.
17. C. Chandler, J. Fairhead, A. Kelly, et al., "Ebola: Limitations of Correcting Misinformation," *Lancet* 385, no. 9975 (2015): 1275–1277.
18. "Together We Can Prevent Ebola," banner for Sierra Leone, Centers for Disease Control and Prevention, 2014, http://www.cdc.gov/vhf/ebola/pdf/bannerforebolasierraleonev2.pdf.
19. "Ebola: A Poem for the Living," animated video, created by United Methodist Communications, October 21, 2014, http://www.comminit.com/ci-ebola/content/ebola-poem-living-video.
20. P. Oosterhoff and A. Wilkinson, "Local Engagement in Ebola Outbreaks and Beyond in Sierra Leone," Practice Paper in Brief 24, Institute of Development Studies, February 2015, https://www.ids.ac.uk/publication/local-engagement-in-ebola-outbreaks-and-beyond-in-sierra-leone.
21. E. T. Richardson, M. B. Barrie, C. Nutt, et al., "The Ebola Suspect's Dilemma," *Lancet* 5, no. 3 (2017): e254–e256.
22. A. Green, "Remembering Health Workers Who Died from Ebola in 2014," *Lancet* 384, no. 9961 (2014): 2201–2206.
23. Green, "Remembering Health Workers."
24. J. Wilson, "Ebola Fears Hit Close to Home," CNN, July 30, 2014, https://www.cnn.com/2014/07/29/health/ebola-outbreak-american-dies/index.html.
25. World Health Organization, "Successful Ebola Responses in Nigeria, Senegal and Mali," January 2015, http://www.who.int/csr/disease/ebola/one-year-report/nigeria/en/.
26. T. Frieden, via Twitter, March 17, 2017, https://twitter.com/DrFrieden/status/842814547442458624.
27. Green, "Remembering Health Workers."
28. D. von Drehle, with A. Baker, "The Ebola Fighters," *Time,* December 10, 2014, http://time.com/time-person-of-the-year-ebola-fighters/.
29. C. Pacheco, "Rethinking the Ebola Response: How Liberians Helped Themselves," Devex media platform, October 5, 2015, https://www

.devex.com/news/rethinking-the-ebola-response-how-liberians-helped-themselves-87030.

30. Adichie, "The Danger of a Single Story."
31. D. M. Secko, B. Morel, and A. Edimo, "An Exploration of the Lived Experience of African Journalists during the 2014 Ebola Crisis," Research Report, 2017, Department of Journalism, Concordia University, Montreal, http://wfsj.org/v2/wp-content/uploads/2017/09/WFSJ_Ebola_Journalism_Research_-Report_FINAL_Sept2017.pdf.
32. World Health Organization, "Health Worker Ebola Infections in Guinea, Liberia and Sierra Leone: Preliminary Report," May 21, 2015, http://www.who.int/csr/resources/publications/ebola/health-worker-infections/en/.
33. Kollie et al., "Deciding to Work."
34. S. Bedrosian, C. Young, L. Smith, et al., "Lessons of Risk Communication and Health Promotion—West Africa and United States," *Morbidity and Mortality Weekly Report* 65, suppl. 3 (2016): 68–74.
35. "Ebola Crisis: Red Cross Says Guinea Aid Workers Face Attacks," BBC News, February 12, 2015, http://www.bbc.com/news/world-africa-31444059.
36. "Bandits in Guinea Steal Blood Samples Believed to Be Infected with Ebola," *Guardian,* November 21, 2014, https://www.theguardian.com/world/2014/nov/21/bandits-guinea-steal-blood-samples-possibly-infected-with-ebola.
37. "Ebola Crisis: Red Cross Says Guinea Aid Workers Face Attacks."
38. J. Mouton, Skype conversation with Salahi, February 9, 2017.
39. K. Sack, S. Fink, P. Belluck, and A. Nossiter, "How Ebola Roared Back," *New York Times,* December 29, 2014.

6. EPIDEMIC OF FEAR

1. Centers for Disease Control and Prevention, "Travelers' Health: Ebola," July 31, 2014, updated May 9, 2017, http://wwwnc.cdc.gov/travel/diseases/ebola.
2. Bhadelia's account is from N. Bhadelia, "Is Stigma an Invisible Killer?" TEDxNatick, March 28, 2016, https://www.youtube.com/watch?v=DR8GOoE91Lk.

3. N. Yozwiak, interview with Salahi, November 15, 2016.
4. Bhadlia, "Is Stigma an Invisible Killer?"
5. Centers for Disease Control and Prevention, "2014–2016 Ebola Outbreak in West Africa," updated December 27, 2017, https://www.cdc.gov/vhf/ebola/history/2014-2016-outbreak/index.html.
6. D. Urbanski, "How Many New Ebola Cases in U.S. by End of 2014? Experts Weigh In," TheBlaze website, November 1, 2014, http://www.theblaze.com/stories/2014/11/01/how-many-new-ebola-cases-in-u-s-by-end-of-2014-experts-weigh-in/.
7. D. Hooper, telephone conversation with Salahi, August 31, 3016.
8. G. Gonsalves and P. Staley, "Panic, Paranoia, and Public Health—The AIDS Epidemic's Lessons for Ebola," *New England Journal of Medicine* 371, no. 25 (2014): 2348–2349.
9. G. Flynn and S. Scutti, "Smuggled Bushmeat Is Ebola's Back Door to America," *Newsweek,* August 21, 2014, http://www.newsweek.com/2014/08/29/smuggled-bushmeat-ebolas-back-door-america-265668.html.
10. G.-L. Whembolua, D. Conserve, D. I. Tshiswaka, et al., "Addressing the Impact of the Ebola Virus Disease–Related Stigma among African Immigrants in the United States," poster presented at the 143rd meeting of the American Public Health Association, November 2, 2015, Chicago.
11. R. Savillo and M. Gertz, "Report: Ebola Coverage on TV News Plummeted after Midterms," Media Matters for America website, November 19, 2014, https://www.mediamatters.org/research/2014/11/19/report-ebola-coverage-on-tv-news-plummeted-afte/201619.
12. G. K. SteelFisher, R. J. Blendon, and N. Lasala-Blanco, "Ebola in the United States—Public Reactions and Implications," *New England Journal of Medicine* 373, no. 9 (2015): 789–791.
13. K. Hills, "Rejecting Quarantine: A Frozen-in-Time Reaction to Disease," in *Ebola's Message: Public Health and Medicine in the Twenty-First Century,* ed. N. G. Evans, T. C. Smith, and M. S. Majumder (Cambridge, MA: MIT Press, 2016), 217, emphasis added.
14. J. M. Drazen, R. Kanapathipillai, E. W. Campion, et al., "Ebola and Quarantine," *New England Journal of Medicine* 371, no. 21 (2014): 2029–2030.
15. Gonsalves and Staley, "Panic, Paranoia, and Public Health."

16. E. Schmall, "Review Faults Dallas Hospital in Ebola Case: Patient with Virus Was Misdiagnosed," *Boston Globe,* September 5, 2015, https://www.bostonglobe.com/news/nation/2015/09/04/review-cites-problems-texas-hospital-during-ebola-crisis/mcbD0jYuZOrCbkpE2UvHUI/story.html. For the review, see "The Expert Panel Report to Texas Health Resources Leadership on the 2014 Ebola Events," available at https://www.calhospital.org/sites/main/files/file-attachments/the_expert_panel_report_to_texas_health_resources.pdf.
17. M. Winter, "Timeline Details Missteps with Ebola Patient Who Died," *USA Today,* October 17, 2014, https://www.usatoday.com/story/news/nation/2014/10/17/ebola-duncan-congress-timeline/17456825/.
18. L. Rosenbaum, "Communicating Uncertainty—Ebola, Public Health, and the Scientific Process," *New England Journal of Medicine* 372, no. 1 (2015): 7–9.
19. Rosenbaum, "Communicating Uncertainty."
20. Schmall, "Review Faults Dallas Hospital."
21. M. G. Kortepeter, P. W. Smith, A. Hewlett, and T. J. Cieslak, "Caring for Patients with Ebola: A Challenge in Any Care Facility," *Annals of Internal Medicine* 162, no. 1 (2015): 68–69.
22. Centers for Disease Control and Prevention, "Hospital Preparedness: A Tiered Approach," last updated February 20, 2015, https://www.cdc.gov/vhf/ebola/healthcare-us/preparing/current-treatment-centers.html.
23. Hills, "Rejecting Quarantine."
24. Rosenbaum, "Communicating Uncertainty."
25. L. M. Garza and T. Wade, "U.S. Health Official Allowed New Ebola Patient on Plane with Slight Fever," Reuters, October 15, 2014, https://www.reuters.com/article/us-health-ebola-usa/u-s-health-official-allowed-new-ebola-patient-on-plane-with-slight-fever-idUSKCN0I40UE20141016.
26. J. Hanrahan, telephone conversation with Salahi, February 20, 2017.
27. J. Hanrahan, "Ebola Preparation in the U.S. before and after Dallas," PowerPoint presentation at the Association of Health Care Journalists annual meeting, April 9, 2016, Cleveland, Ohio.
28. J. Hanrahan, "Ebola Preparation in the U.S. before and after Dallas."
29. J. Hanrahan, telephone conversation with Salahi, February 20, 2017.

30. J. Hanrahan, telephone conversation with Salahi, February 20, 2017.
31. C. Fraser, S. Riley, R. M. Anderson, and N. M. Ferguson, "Factors That Make an Infectious Disease Outbreak Controllable," *Proceedings of the National Academy of Sciences* 101, no. 16 (2004): 6146–6151.
32. Hills, "Rejecting Quarantine."
33. T. Svoboda, B. Henry, L. Shulman, et al., "Public Health Measures to Control the Spread of Severe Acute Respiratory Syndrome during the Outbreak in Toronto," *New England Journal of Medicine* 350, no. 23 (2004): 2352–2361.
34. "Kaci Hickox's Settlement with Christie Administration Creates Quarantine 'Bill of Rights,'" July 27, 2017, ACLU of New Jersey, https://www.aclu-nj.org/news/2017/07/27/victory-detained-nurses-ebola-suit-secures-due-process.
35. D. G. McNeil Jr., "Cuba's Fortresses against a Viral Foe," *New York Times,* May 7, 2012, http://www.nytimes.com/2012/05/08/health/cubas-aids-sanitariums-fortresses-against-a-viral-foe.html?mtrref=undefined.
36. S. Z. Hoffman, "HIV/AIDS in Cuba: A Model for Care or an Ethical Dilemma?" *African Health Sciences* 4, no. 3 (2004): 208–209.
37. Nancy Snyderman, as told to Seth Abramovitch, "Nancy Snyderman Breaks Silence on Ebola Nightmare, NBC News: 'People Wanted Me Dead,'" August 26, 2015, Hollywood Reporter online, http://www.hollywoodreporter.com/features/nancy-snyderman-breaks-silence-ebola-817601.
38. Quoted in H. Lewis, "NBC News' Nancy Snyderman Admits She Violated Voluntary Ebola Quarantine, Apologizes," December 3, 2014, Hollywood Reporter online, http://www.hollywoodreporter.com/news/nbc-news-nancy-snyderman-admits-753594.
39. SteelFisher et al., "Ebola in the United States."
40. "Majority of Americans Believe Ebola Spreads through Air: Poll," Reuters, October 15, 2014, https://www.reuters.com/article/us-health-ebola-usa-poll/majority-of-americans-believe-ebola-spreads-through-air-poll-idUSKCN0I42JA20141015; World Health Organization, "What We Know about Transmission of the Ebola Virus among Humans," October 6, 2014, http://www.who.int/mediacentre/news/ebola/06-october-2014/en/.
41. SteelFisher et al., "Ebola in the United States."

7. INVESTMENT AND ACCOUNTABILITY

1. N. Bhadelia, telephone conversation with Salahi, April 5, 2016.
2. "Ebola Outbreak: Sierra Leone Workers Dump bodies in Kenema," BBC, November 25, 2014, https://www.bbc.com/news/world-africa-30191938.
3. A. Maxmen, "In Fight against Ebola, Front-Line Health Workers Risked Their Lives and Never Got Paid," *Newsweek,* May 19, 2015, http://www.newsweek.com/2015/06/05/fight-against-ebola-front-line-health-workers-were-sidelined-funding-333436.html.
4. N. Bhadelia, telephone interviews with Salahi on March 28, 2016, and April 5, 2016, and in-person interview with Salahi on April 14, 2017.
5. N. Bhadelia, "Health Security and Emerging Infections" (lecture, Fletcher Ideas Exchange, Fletcher School of Diplomacy, Tufts University, Medford, MA, April 14, 2016), https://www.youtube.com/watch?v=yY0y0CuK1Q0.
6. International Federation of Red Cross and Red Crescent Societies, "IFRC Statement on Fraud in Ebola Operations," October 20, 2017, https://media.ifrc.org/ifrc/ifrc-statement-fraud-ebola-operations/.
7. International Federation of Red Cross and Red Crescent Societies, "IFRC Statement on Fraud in Ebola Operations."
8. S. Moon, J. Leigh, L. Woskie, et al., "Post-Ebola Reforms: Ample Analysis, Inadequate Action," *BMJ* 356 (2017): j280, doi: 10.1136/bmj.j280.
9. Save the Children, "A Wake-up Call: Lessons from Ebola for the World's Health Systems" (London: Save the Children, 2015), https://www.savethechildren.org.uk/content/dam/global/reports/health-and-nutrition/a-wake-up-call.pdf.; Chatham House, "Shared Responsibilities for Health: A Coherent Global Framework for Health Financing," Chatham House Report, May 2014, https://www.chathamhouse.org/sites/files/chathamhouse/field/field_document/20140521HealthFinancing.pdf.
10. F. Poquie, "Liberia Seeks $1.3 Billion to Revive Economy after Ebola," Bloomberg website, October 31, 2016, https://www.bloomberg.com/news/articles/2016-11-01/liberia-seeks-1-3-billion-to-revive-economy-after-ebola.
11. Save the Children, "Wake-up Call."

12. International Ebola Recovery Conference, Summary Report, July 20, 2015, https://ebolaresponse.un.org/sites/default/files/summary_report.pdf.

13. J. Kates, J. Michaud, A. Wexler, and A. Valentine, "The U.S. Response to Ebola: Status of the FY2015 Emergency Ebola Appropriation," Issue Brief, Henry J. Kaiser Family Foundation, December 11, 2015, https://www.kff.org/global-health-policy/issue-brief/the-u-s-response-to-ebola-status-of-the-fy2015-emergency-ebola-appropriation/.

14. White House, "Emergency Funding Request to Enhance the U.S. Government's Response to Ebola at Home and Abroad," Fact Sheet, Office of the Press Secretary, November 5, 2014, https://obamawhitehouse.archives.gov/the-press-office/2014/11/05/fact-sheet-emergency-funding-request-enhance-us-government-s-response-eb.

15. Kates et al., "U.S. Response to Ebola."

16. V. Ramachandran and J. Walz, "Haiti: Where Has All the money Gone?" Center for Global Development, Policy Paper 004, May 2012, https://www.cgdev.org/sites/default/files/1426185_file_Ramachandran_Walz_haiti_FINAL_0.pdf.

17. Audit Service Sierra Leone, "Report on the Audit of the Management of the Ebola Funds: May to October 2014," Freetown, Sierra Leone, October 31, 2014, http://www.auditservice.gov.sl/report/assl-report-on-ebola-funds-management-may-oct-2014.pdf.

18. Audit Service Sierra Leone, "Report on the Audit."

19. L. Taylor-Pearce (Auditor General of Sierra Leone), interviewed by Jamie Hitchen on transparency and accountability in public financial management, April 29 and August 4, 2016, Africa Research Institute, November 9, 2016, http://www.africaresearchinstitute.org/newsite/wp-content/uploads/2016/11/ARI-Conversations-Series-LaraTaylor-Pearce-3.pdf.

20. M. Massaquoi, "Health Minister Wants World Bank Ebola Money Probe," *Concord Times* (Sierra Leone), May 14, 2015, http://slconcordtimes.com/health-minister-wants-world-bank-ebola-money-probe/.

21. World Health Organization, The WHO Programme Budget Portal, http://open.who.int/2018-19/home.

22. B. Friel, telephone conversation with Salahi, October 31, 2016.

23. J. Wu, "Experts Cite Ebola's Indirect Cost, Urge Public-Private Partnerships," Medill Reports Chicago, Medill School of Journalism, Northwestern University, January 26, 2015, http://news.medill.northwestern.edu/chicago/experts-cite-ebolas-indirect-cost-urge-public-private-partnerships/.

24. US Department of State contract for Emergency Aeromedical Evacuation Services, August 18, 2014, https://www.fbo.gov/index?s=opportunity&mode=form&id=dede7398603a794323675588fbd4609e&tab=core&_cview=0.

25. US Department of Defense, Defense Procurement and Acquisition Policy, "Justifications and Approvals / Sole Source Acquisitions," page last updated December 23, 2014, http://www.acq.osd.mil/dpap/ccap/cc/jcchb/HTML/Topical/sole_source.html.

26. J. Fair, telephone interview with Salahi, July 7, 2017.

27. Medill National Security Reporting Project, "This US Government Program May Have Stopped Ebola—But It Never Had the Funding It Requested," VICE News, July 8, 2016, https://news.vice.com/article/this-us-government-program-may-have-stopped-ebola-but-never-had-the-funding-it-requested.

28. Boston University National Emerging Infectious Diseases Laboratory, "EIDA2Z Symposium," news release, https://www.bu.edu/neidl/news/eida2z/.

29. T. Howell Jr., "Obama Should Use $2.7B Ebola Leftover Funds to Fight Zika," *Washington Times,* February 18, 2016, http://www.washingtontimes.com/news/2016/feb/18/zika-virus-funds-should-be-diverted-from-ebola-res/.

30. A. Kodjak, "Congress Ends Spat, Agrees to Fund $1.1 billion to Combat Zika," September 28, 2016, NPR, https://www.npr.org/sections/health-shots/2016/09/28/495806979/congress-ends-spat-over-zika-funding-approves-1-1-billion.

31. D. Scott, "Millions in Ebola Funding, a Casualty of Zika Virus, May Not Be Replenished," STAT news, June 1, 2016, https://www.statnews.com/2016/06/01/ebola-zika-virus-funding/.

32. Medill National Security Reporting Project, "This US Government Program May Have Stopped Ebola."

33. World Bank, "The Economic Impact of Ebola on Sub-Saharan Africa: Updated Estimates for 2015," Working Paper no. 93721, World Bank

Group, Washington, DC, January 20, 2015, http://documents.worldbank.org/curated/en/541991468001792719/pdf/937210REVISED000Jan02002015000FINAL.pdf.

34. World Bank, "The Economic Impact of the 2014 Ebola Epidemic: Short and Medium Term Estimates for West Africa," Working Paper no. 91219, World Bank Group, Washington, DC, October 7, 2014, http://documents.worldbank.org/curated/en/524521468141287875/pdf/912190WP0see0a00070385314B00PUBLIC0.pdf.

35. Moon et al., "Post-Ebola Reforms."

36. World Bank, "World Bank Group Launches Groundbreaking Financing Facility to Protect Poorest Countries against Pandemics," press release, May 21, 2016, http://www.worldbank.org/en/news/press-release/2016/05/21/world-bank-group-launches-groundbreaking-financing-facility-to-protect-poorest-countries-against-pandemics.

37. M. Osterholm, telephone interview with Salahi, January 24, 2018.

38. T. W. Geisbert, "First Ebola Virus Vaccine to Protect Human Beings?" *Lancet* 389, no. 10068 (2017): 479–480.

39. N. D. Gunaratne, "The Ebola Virus and the Threat of Bioterrorism," *Fletcher Forum of World Affairs* 39, no. 1 (2015): 63–76; Request for Information, "Ebola within a Bioterrorism Context," October 2, 2014, https://assets.publishing.service.gov.uk/government/uploads/system/uploads/attachment_data/file/399328/Attachment_Ebola_within_a_bioterrorism_context.pdf, retrieved from [UK] Ministry of Defence, Freedom of Information responses released during the week commencing 26 January 2015, https://www.gov.uk/government/publications/foi-responses-released-by-mod-week-commencing-26-january-2015. A letter concerning the FOI response, from the Defence Science and Technology Strategy Secretariat, London, January 23, 2015, is at https://www.gov.uk/government/uploads/system/uploads/attachment_data/file/399322/Ebola_within_a_bioterrorism_context.pdf.

40. US Department of Health and Human Services, Project BioShield Overview, MedicalCounterMeasures.gov, last updated October 18, 2016, https://www.medicalcountermeasures.gov/barda/cbrn/project-bioshield-overview/.

8. EBOLA'S FALLOUT

Epigraph: C. E. Rosenberg, "What Is an Epidemic? AIDS in Historical Perspective," *Daedalus* 118, no. 2 (1989): 1–17.

1. World Health Organization, "Latest Outbreak Over in Liberia; West Africa Is at Zero, but New Flare-ups Are Likely to Occur," World Health Organization News Release, January 14, 2016, http://www.who.int/mediacentre/news/releases/2016/ebola-zero-liberia/en/.
2. L. Schlein, "W. Africa Better Prepared to Contain Future Ebola Outbreaks," VOA, May 29, 2016, https://www.voanews.com/a/who-official-liberia-sierra-leone-guinea-skillfully-managed-ebola-flare-ups/3350650.html.
3. J. Liu, "Disease Outbreak: Finish the Fight against Ebola," *Nature* 524, no. 7563 (2015): 27–29.
4. Liu, "Disease Outbreak."
5. World Health Organization, Global Health Observatory country views, Guinea statistics summary (2002–present), last updated April 3, 2018, http://apps.who.int/gho/data/node.country.country-GIN.
6. E. C. Hayden, "Ebola's Lasting Legacy," *Nature* 519, no. 7541 (2015): 24–26.
7. Liu, "Disease Outbreak."
8. J. J. Farrar and P. Piot, "The Ebola Emergency—Immediate Action, Ongoing Strategy," *New England Journal of Medicine* 371, no. 16 (2014): 1545–1546.
9. A. S. Parpia, M. L. Ndeffo-Mbah, N. S. Wenzel, and A. P. Galvani, "Effects of Response to 2014–2015 Ebola Outbreak on Deaths from Malaria, HIV / AIDS, and Tuberculosis, West Africa," *Emerging Infectious Diseases* 22, no. 3 (2016): 433–441.
10. R. S. Dhillon and J. D. Kelly, "Community Trust and the Ebola Endgame," *New England Journal of Medicine* 373, no. 9 (2015): 787–789.
11. Liu, "Disease Outbreak."
12. D. Jallah, email communication with Salahi, May 10, 2017.
13. "Rebuilding Health Care in the Shadow of Ebola," The Global Fund, United Nations Office for the Coordination of Humanitarian Affairs, February 20, 2017, https://reliefweb.int/report/sierra-leone/rebuilding-health-care-shadow-ebola; A. Green, "West African Countries Focus on Post-Ebola Recovery Plans," *Lancet* 388, no. 10059 (2016):

2463–2465; D. K. Evans, M. Goldstein, and A. Popova, "Health-Care Worker Mortality and the Legacy of the Ebola Epidemic," Correspondence, *Lancet Global Health* 3, no. 8 (2015): e439–e440.

14. P. Vetter, L. Kaiser, M. Schibler, et al., "Sequelae of Ebola Virus Disease: The Emergency within an Emergency," *Lancet Infectious Diseases* 16, no. 6 (2016): e82–e91.

15. J. T. Scott, F. R. Sesay, T. A. Massaquoi, et al., "Post-Ebola Syndrome, Sierra Leone," *Emerging Infectious Diseases* 22, no. 4 (2016): 641–646.

16. Partners in Health, "Blindness," Partners in Health website, January 9, 2017, http://www.pih.org/blog/blindness-uveitis-partners-in-health-sierra-leone-eye-care.

17. "Ebola Victims Sue Sierra Leone Government over Mismanaged Funds," Reuters, December 15, 2017, https://www.reuters.com/article/us-health-ebola-leone/ebola-victims-sue-sierra-leone-government-over-mismanaged-funds-idUSKBN1E92NE.

18. World Health Organization, "Ebola Situation Report—20 January 2016," http://apps.who.int/ebola/current-situation/ebola-situation-report-20-january-2016.

19. "West Africa—Ebola Outbreak," Fact Sheet #12, Fiscal Year 2016, USAID/CDC, September 30, 2016, https://www.usaid.gov/sites/default/files/documents/1866/west_africa_ebola_fs12_09-30-2016.pdf.

20. National Institutes of Health, "Study of Ebola Survivors Opens in Liberia," NIH News Release, June 17, 2015, https://www.nih.gov/news-events/news-releases/study-ebola-survivors-opens-liberia.

21. J. M. Epstein, L. M. Sauer, J. Chelen, et al., "Infectious Disease: Mobilizing Ebola Survivors to Curb the Epidemic," *Nature* 516, no. 7531 (2014): 323–325.

22. Green, "West African Countries."

23. L. H. Sun, "Outgoing CDC Chief Talks about Agency's Successes and His Greatest Fear," *Washington Post,* January 16, 2017, https://www.washingtonpost.com/news/to-your-health/wp/2017/01/16/outgoing-cdc-chief-talks-about-the-agencys-successes-and-his-greatest-fear/?utm_term=.df2589979b8c.

24. L. Rosenbaum, "Communicating Uncertainty—Ebola, Public Health, and the Scientific Process," *New England Journal of Medicine* 371, no. 1 (2015): 7–9.

25. WHO Ebola Response Team, "Ebola Virus Disease in West Africa—The First 9 Months of the Epidemic and Forward Projections," *New England Journal of Medicine* 371, no. 16 (2014): 1481–1495.

26. Partnership for Research on Ebola in Liberia (PREVAIL) II Writing Group, "A Randomized, Controlled Trial of ZMapp for Ebola Virus Infection," *New England Journal of Medicine* 375, no. 15 (2016): 1448–1456.

27. K. Huster, "Ebola: We May Have Won the Battle but We Haven't Won the War," KPBS, NPR, March 27, 2016, http://www.kpbs.org/news/2016/mar/27/ebola-we-may-have-won-the-battle-but-we-havent/?amp=amp.

28. Global Health Security Agenda, "About," GHSA website, https://www.ghsagenda.org/about.

29. A. Fauci and V. Kerry, "Preventing a Pandemic" (presentation at PULSE: On the Front Lines of Health Care forum, Boston, MA, June 13, 2017).

30. E. C. Hayden, "Spectre of Ebola Haunts Zika Response," *Nature* 531, no. 7592 (2016): 19.

31. C. Zimmer, *A Planet of Viruses,* 2nd ed. (Chicago: University of Chicago Press, 2015).

32. E. C. Holmes, A. Rambaut, and K. G. Andersen, "Pandemics: Spend on Surveillance, Not Prediction," *Nature* 580, no. 7709 (2018): 180–182.

33. K. F. Smith, M. Goldberg, S. Rosenthal, et al., "Global Rise in Human Infectious Disease Outbreaks," *Journal of the Royal Society Interface* 11 (2014): 20140950.

34. J. L. Geoghegan, A. M. Senior, F. DiGiallonardo, and E. C. Holmes, "Virological Factors That Increase the Transmissibility of Emerging Human Viruses," *Proceedings of the National Academy of Sciences* 113, no. 15 (2016): 4170–4175.

35. L. H. Taylor, S. M. Latham, and M. E. Woolhouse, "Risk Factors for Human Disease Emergence," *Philosophical Transactions of the Royal Society of London, B: Biological Sciences* 356, no. 1411 (2001): 983–989; Centers for Disease Control and Prevention, "Zoonotic Diseases," CDC One Health Basics, last updated April 2, 2018, https://www.cdc.gov/onehealth/basics/zoonotic-diseases.html.

36. World Health Organization, "Factors That Contributed to Undetected Spread of the Ebola Virus and Impeded Rapid Containment,"

WHO Emergency Preparedness, Response, January 2015, http://www.who.int/csr/disease/ebola/one-year-report/factors/en/.

9. NAVIGATING THE NEXT EPIDEMIC

1. GHRF Commission (Commission on a Global Health Risk Framework for the Future), "The Case for Investing in Pandemic Preparedness," in *The Neglected Dimension of Global Security: A Framework to Counter Infectious Disease Crises,* National Academy of Medicine (Washington, DC: National Academies Press, 2016), chap. 2, https://www.ncbi.nlm.nih.gov/books/NBK368391/.
2. S. Moon, D. Sridhar, M. A. Pate, et al., "Will Ebola Change the Game? Ten Essential Reforms before the Next Pandemic. The Report of the Harvard-LSHTM Independent Panel on the Global Response to Ebola," *Lancet* 386, no. 10009 (2015): 2204–2221.
3. L. O. Gostin, J. W. Sapsin, S. P. Teret, et al., "The Model State Emergency Health Powers Act: Planning for and Response to Bioterrorism and Naturally Occurring Infectious Diseases," *Journal of the American Medical Association* 288, no. 5 (2002): 622–628.
4. P. Piot, telephone conversation with Sabeti and Salahi, April 6, 2017.
5. L. S. Graybill, "Traditional Practices and Reconciliation in Sierra Leone: The Effectiveness of Fambul Tok," *Conflict Trends* 2010, no. 3, (2010): 41–47. See also Chapter 1.
6. S. Galea and G. Annas, "An Argument for a Common-Sense Global Public Health Agenda," *Lancet Public Health* 2, no. 10 (2017): e445–e446.
7. A. Fauci and V. Kerry, "Preventing a Pandemic" (presentation at PULSE: On the Front Lines of Health Care forum, Boston, MA, June 13, 2017).
8. J. Farrar, telephone conversation with Salahi, April 28, 2017.
9. General Accounting Office, Defense Civil Support: *DoD, HHS, and DHS Should Use Existing Coordination Mechanisms to Improve Their Pandemic Preparedness,* GAO-17-150 (Washington, DC: Government Accountability Office, 2017), https://www.gao.gov/assets/690/682707.pdf.
10. M. Chan, "Report by the Director-General to the Special Session of the Executive Board on Ebola," World Health Organization, January 25,

2015, http://www.who.int/dg/speeches/2015/executive-board-ebola/en/.

11. D. H. Peters, G. T. Keusch, J. Cooper, et al., "In Search of Global Governance for Research in Epidemics," *Lancet* 390, no. 10103 (2017): 1632–1633.
12. World Health Organization, *An R&D Blueprint for Action to Prevent Epidemics: Plan of Action* (Geneva: World Health Organization, 2016). For general information and an annually updated list of priority diseases, see http://www.who.int/blueprint/about/en/.
13. A. M. Henao-Restrepo, A. Camacho, I. M. Longini, et al., "Efficacy and Effectiveness of an rVSV-Vectored Vaccine in Preventing Ebola Virus Disease: Final Results from the Guinea Ring Vaccination, Open-Label, Cluster Randomised Trial (Ebola Ça Suffit!)," *Lancet* 389, no. 10068 (2017): 505–518.
14. Peters et al., "In Search of Global Governance."
15. GHRF Commission, "Building a Framework for Global Security," in *Neglected Dimension of Global Security*, chap. 6, https://www.ncbi.nlm.nih.gov/books/NBK368389/.
16. "World Military Spending: Increases in the USA and Europe, Decreases in Oil-Exporting Countries," Stockholm International Peace Research Institute website, April 24, 2017, https://www.sipri.org/media/press-release/2017/world-military-spending-increases-usa-and-europe.
17. J. J. Farrar and P. Piot, "The Ebola Emergency—Immediate Action, Ongoing Strategy," *New England Journal of Medicine* 371, no. 16 (2014): 1545–1546.
18. "President Donald J. Trump Is Protecting American Taxpayer Dollars," White House Briefing, May 8, 2018, https://www.whitehouse.gov/briefings-statements/president-donald-j-trump-protecting-american-taxpayer-dollars/; "Congo Declares Ebola Outbreak after 17 People Die From Virus," *Bloomberg,* May 9, 2018, http://fortune.com/2018/05/09/congo-ebola-virus-outbreak/.
19. Centers for Disease Control and Prevention, "Global Health—CDC and the Global Health Security Agenda," https://www.cdc.gov/globalhealth/security/index.htm; E. Baumgaertner, "White House Hails Success of Disease-Fighting Program, and Plans Deep Cuts," *New York Times,* March 13, 2018, https://www.nytimes.com/2018/03/13/us/politics/trump-ebola-disease-cuts-global-health-security-agenda.html.

20. P. Roddy, D. Weatherhill, B. Jeffs, et al., "The Médecins sans Frontières Intervention in the Marburg Hemorrhagic Fever Epidemic, Uige, Angola, 2005. II. Lessons Learned in the Community," *Journal of Infectious Diseases* 196 (2007): S162–S167.
21. T. M. Larsen, C. B. Mburu, A. Kongelf, et al., "Red Cross Volunteers' Experience with a Mobile Community Event-Based Surveillance (CEBS) System in Sierra Leone during and after the Ebola Outbreak—A Qualitative Study," *Health and Primary Care* 1, no. 3 (2017): 1–7.
22. P. M. Sandman and J. Lanard, "Explaining and Proclaiming Uncertainty: Risk Communication Lessons from Germany's Deadly *E. coli* Outbreak," Peter Sandman website, posted August 4, 2011, http://www.psandman.com/col/GermanEcoli.htm.
23. G. K. SteelFisher, R. J. Blendon, and N. Lasala-Blanco, "Ebola in the United States—Public Reactions and Implications," *New England Journal of Medicine* 373, no. 9 (2015): 789–791.
24. D. L. Heymann, D. Barakamfitiye, M. Szczeniowski, et al., "Ebola Hemorrhagic Fever: Lessons from Kikwit, Democratic Republic of the Congo," *Journal of Infectious Diseases* 179, suppl. 1 (1999): S283–S286.

EPILOGUE

1. "Flagship Genome Research Centre for Africa," *African Design Magazine* 38 (March 2018): 24–30.

ACKNOWLEDGMENTS

In the name of God, the Most Merciful, Most Compassionate. Our deepest gratitude goes to all who sacrificed their lives to help others during the Ebola epidemic. Thank you to the Khan family for sharing with us a candid account of the beloved Dr. Sheik Humarr Khan (God have mercy on his soul) and trusting us to amplify his life's contribution to the world. Dr. Khan was a fighter for the people of Sierra Leone, and his legacy is forever sewn into the fabric of the country. The staff at Kenema Government Hospital are the bravest and most compassionate people we know. We are grateful to them for speaking their truth with such openness.

Thank you to Debbie Wilson for telling her story and introducing us to healthcare workers at Liberia's Foya Hospital. Darlington Jallah consistently reminded us that every victim, survivor, family member, worker, and responder has a name.

Those who took the time to complete our survey in the aftermath of a traumatic event helped us capture the emotions and resilience of infectious disease outbreak responders. Their honest accounts forced us to dig deeper into what the world had experienced and gave this book its legs. We would also like to acknowledge all of the public health and global health leaders, experts, scientists, clinicians, and executives who unfailingly provided the research material, anecdotes, and assurances that we needed to write this book. Only some are named, but all offered diverse perspectives, dissuaded us from single narratives, and challenged us to seek truth. Among them were Amesh Adalja, Bruce Aylward, Richard Besser, Nahid Bhadelia, Sylvia Blyden, Sheila Davis, Joseph Fair, Paul Farmer, Jeremy Farrar, Jonathan Fielding, Brian Friel, Bob Garry, Steven Gordon, Jennifer Hanrahan, Christian Happi, David Hooper, Simbirie Jalloh, Jerome Mouton, Michael Osterholm, Peter Piot, Robert Quigley, Indi Trehan, and Annie Wilkinson.

We are forever grateful to all who have worked and still work at the Sabeti Lab for being so welcoming of our idea of conducting such research and for being so willing to allow a reporter in for candid inquiries. You made her feel like one of you. Kayla Barnes, Michael Butts, August Felix, Stephen Gire, Ashley Matthews, Nathan Yozwiak, among others, all in their own ways added insight, value, and organization to this undertaking. Our undergraduate researcher, Rushabh Doshi, took on many roles, including data analyzer, graphic designer, fact checker, and formatter. He never missed a meeting, even if it meant catching a red-eye to make a morning start. He has already accomplished so much and is destined for greater things.

It did not take much convincing for Harvard University Press to recognize that the central themes of this book transcend any one epidemic in any one region. Janice Audet championed the project from conception to publication. Louise Robbins carried it to completion. We are grateful to have found a home for our book with HUP.

We are grateful to Lara Setrakian for connecting us and igniting the path for this collaborative endeavor. We are thankful to our families for believing in us, believing in our work, and believing in this book. To David, who supports every big idea, and to Dalia and Adam, to whom the world and all of its beauty and discovery belong—the pages of a book cannot hold the amount of gratitude we feel. Finally, we are indebted to our parents for raising us in a country that gave us the freedom and opportunities to explore, discover, reach, and realize.

INDEX

A page number followed by *t* indicates a table. An *italicized* page number indicates a photograph.